Lecture Notes
in Business Information Processing   **351**

Series Editors

Wil van der Aalst
RWTH Aachen University, Aachen, Germany
John Mylopoulos
University of Trento, Trento, Italy
Michael Rosemann
Queensland University of Technology, Brisbane, QLD, Australia
Michael J. Shaw
University of Illinois, Urbana-Champaign, IL, USA
Clemens Szyperski
Microsoft Research, Redmond, WA, USA

More information about this series at http://www.springer.com/series/7911

Danielle Costa Morais ·
Ashley Carreras · Adiel Teixeira de Almeida ·
Rudolf Vetschera (Eds.)

Group Decision and Negotiation

Behavior, Models, and Support

19th International Conference, GDN 2019
Loughborough, UK, June 11–15, 2019
Proceedings

 Springer

Editors
Danielle Costa Morais 🆔
CDSID - Center for Decision Systems
and Information Development
Universidade Federal de Pernambuco
(UFPE)
Recife, Brazil

Adiel Teixeira de Almeida 🆔
CDSID - Center for Decision Systems
and Information Development
Universidade Federal de Pernambuco
(UFPE)
Recife, Brazil

Ashley Carreras 🆔
Loughborough University
Loughborough, UK

Rudolf Vetschera 🆔
University of Vienna
Vienna, Austria

ISSN 1865-1348   ISSN 1865-1356 (electronic)
Lecture Notes in Business Information Processing
ISBN 978-3-030-21710-5   ISBN 978-3-030-21711-2 (eBook)
https://doi.org/10.1007/978-3-030-21711-2

This Springer imprint is published by the registered company Springer Nature Switzerland AG
The registered company address is: Gewerbestrasse 11, 6330 Cham, Switzerland

Preface

The annual conferences on Group Decision and Negotiation have become an important meeting point for researchers interested in the many aspects of collective decision-making. What started out as a one-time event at the beginning of the millennium has developed into a series of conferences that have been held (with one exception) every year since 2000. GDN is a truly global conference uniting researchers from all over the world, which up to now has been held in five continents: once each in Australia (Perth 2002), South America (Recife 2012), and Asia (Nanjing 2018), four times in North America (Banff 2004, Mt. Tremblant 2007 and Toronto 2009, all in Canada, and Bellingham, USA 2016), and 11 times in Europe (Glasgow 2000, La Rochelle 2001, Istanbul 2003, Vienna 2005, Karlsruhe 2006, Coimbra 2008, Delft 2010, Stockholm 2013, Toulouse 2014, Warsaw 2015, and Stuttgart 2017).

In 2019, GDN returned to Europe. We are very grateful to Loughborough University for hosting the first joint GDN and Behavioral OR conference. This joint meeting of the INFORMS/GDN Section and EURO Working Group Behavioral OR brought together two closely connected, but still distinct, communities.

In total, 98 papers grouped into nine different streams were submitted for the conference, covering a wide range of topics related to group decisions, negotiations, and behavioral OR. The largest streams were BOR (22 papers), Preference Modeling for GDN (14 papers) and Collaborative Decision-Making Processes (14 papers). After a thorough review process, 17 of these papers were selected for inclusion in this volume entitled *Group Decision and Negotiation: Behavior, Models and Support*. We have organized the volume according to the five main streams of the conference: "Preference Modeling for GDN," "Collaborative Decision-Making Processes," "Conflict Resolution," "Behavioral Operations Research," and "Negotiation Support Systems and Studies (NS3)."

The section "Preference Modeling for Group Decision and Negotiation" contains five papers that present different models in various contexts. Frej, de Almeida, and Roselli in their paper "Solving Multicriteria Group Decision-Making (MCGDM) Problems Based on Ranking with Partial Information" present a group decision process based on the FITradeoff method. They propose improved graphical visualization features based on behavioral studies using neuroscience tools. The second paper by Luo, "How to Deal with the Multiple Sources of Influences in Group Decision-Making? From a Nonordering to an Ordering Approach," extends the classic social choice functions to signed and weighted social influence functions. In the following paper, Aslan, Dindar, and Laine investigate the properties of seat-wise majority voting to choose a committee in their paper "Choosing a Committee Under Majority Voting." The paper "Reciprocity and Rule Preferences of a Rotating Savings and Credit Association (ROSCA) in China: Evolutionary Simulation in Imitation Games" by Zhao and Horita analyzes how peoples' preference for a ROSCA are related to the reciprocity level in a particular society. The last paper of this section, by Leoneti and

Ziotti, entitled "Modeling the Conflict Within Group Decision-Making: A Comparison Between Methods That Require and do not Require the Use of Preference Aggregation Techniques" presents a comparison of MCDM methods for their application within a group decision-making process with respect to the preference aggregation procedures.

The second section is related to "Collaborative Decision-Making Processes" and contains four papers. Sakka, Bosetti, Grigera, Camilleri, Fernández, Zarat, Bismonte, and Sautot, in their paper "UX Challenges in GDSS: An Experience Report" present three user tests of a collaborative framework called GRoUp Support (GRUS) conducted in three different countries. In the second paper "A Voting Procedures Recommender System for Decision-Making." Coulibaly, Zaraté, Camilleri, Konate, and Tangara propose the use of GRECO (Group vote RECOmmendation), which currently implements Borda, Condorcet, plurality, Black and Copeland voting procedures. Another important aspect in decision-making processes is the generation of the objectives. This topic is considered by Ferretti in the paper "Why Is it Worth it to Expand Your Set of Objectives? Impacts from Behavioral Decision Analysis in Action," where two real interventions of value-focused thinking to support both private and public organizations in generating objectives within strategic decision-making processes are described. The last paper of this section by Tseng and Kou, entitled "Identifying and Ranking Critical Success Factors for Implementing Financial Education in Taiwan Elementary Schools" presents the results of a two-round Delphi survey to identify the critical success factors for successfully implementing financial education.

In the section on "Conflict Resolution," three papers of different contexts are presented. Zeleznikow and Prawer discuss the role of armed international conflict in their paper "War as a Technique of International Conflict Resolution – An Analytical Approach." Fang, Xu, Perc, and Chen in the paper entitled "The Effect of Conformists' Behavior on Cooperation in the Spatial Public Goods Game" studied social dilemmas, investigating the effects of rational and irrational conformity behavior on the evolution of cooperation in a public goods game. Abraham and Ramachandran present the paper "Effect of Pollution on Transboundary River Water Trade," which studies the impact of pollution on river water allocation between riparian states during conflict and cooperation.

The next section contains three papers related to "Behavioral Operations Research (BOR)." The first one by Roszkowska and Wachowicz, entitled "Cognitive Style and the Expectations Toward the Preference Representation in Decision Support Systems" presents a study of the decision-makers' expectations regarding the results in DSS (e.g., ranking vs. ratings) when decision-makers can express their preferences in different ways (numbers and words). The second paper deals with anger in e-negotiation. In their paper "Cue Usage Characteristics of Angry Negotiators in Distributive Electronic Negotiation," Venkiteswaran and Sundarraj conducted an experiment for analyzing the usage of cues (statements and para-linguistic cues including emoticons) by angry negotiators. The last paper of this section is by Ishii, entitled "Opinion Dynamics Theory Considering Trust and Suspicion in Human Relations." This paper discusses trust and distrust among people in a society, and specifically considers mass media effects.

The final section "Negotiation Support Systems and Studies (NS3)" contains two papers describing new developments in NSSs. Schmid and Schoop, in their paper "A Framework for Gamified Electronic Negotiation Training," present a novel approach for e-negotiation trainings by including game elements in an NSS. A study of the use of pattern recognition in e-negotiation data for descriptive and predictive tasks is presented by Kaya and Schoop in the paper "Application of Data Mining Methods for Pattern Recognition in Negotiation Support Systems."

The preparation of the conference and of this volume required the efforts and collaboration of many people. In particular, we thank the general chair of GDN 2019, Marc Kilgour, for his continuous contribution for the GDN Section. Special thanks also go to all the stream chairs: Raimo Hämäläinen, Luis Alberto Franco (BOR), Liping Fang, Keith W. Hipel, D. Marc Kilgour (Conflict Resolution), Tomasz Wachowicz (Preference Modeling for Group Decision and Negotiation), Pascale Zarperformers (Collaborative Decision Making Processes), Bilyana Martinovski (Emotion in Group Decision and Negotiation), Xusen Cheng, G.J. de Vreede (Crowdsourcing), Haiyan Xu, Shawei He (Risk Evaluation and Negotiation Strategies), Mareike Schoop, and Philipp Melzer (Negotiation Support Systems and Studies - NS3). We also thank the reviewers for their timely and informative reviews: Ana Paula Costa, Andreas Schmid, Annika Lenz, Ayşegül Engin, Barbara Göbl, Bilyana Martinovski, Bogumil Kaminski, Christian Stummer, Colin Williams, Dmitry Gimon, Ewa Roszkowska, Ginger Ke, Hannu Nurmi, Ilkka Leppanen, Jing Ma, Kevin Li, Leandro C. Rego, Lihi Naamani-Dery, Love Ekenberg, Maisa Silva, Marc Fernandes, Marc Kilgour, Marcella Urtiga, Masahide Horita, Pascale Zarperformers, Patrick Buckley, Per van der Wijst, Peter Kesting, Philipp Melzer, Przemyslaw Szufel, Raimo Hämäläinen, Rustam Vahidov, Sabine Koeszegi, Shikui Wu, Simone Philpot, Tobias Langenegger, Tomasz Szapiro, Tomasz Wachowicz, Will Baber, and Yi Xiao.

We also are very grateful to Ralf Gerstner, Alfred Hofmann, and Christine Reiss at Springer for the excellent collaboration.

April 2019

Danielle Costa Morais
Ashley Carreras
Adiel Teixeira de Almeida
Rudolf Vetschera

Organization

Honorary Chair

Marc Kilgour   Wilfrid Laurier University, Canada

General Chairs

Alberto Franco   Loughborough University, UK
Rudolf Vetschera   University of Vienna, Austria

Program Chairs

Danielle Costa Morais   Universidade Federal de Pernambuco, Brazil
Ashley Carreras   Loughborough University, UK
Adiel Teixeira de Almeida   Universidade Federal de Pernambuco, Brazil

Program Committee

Amer Obeidi	University of Waterloo, Canada
Bilyana Martinovski	Stockholm University, Sweden
Bo Yu	Dalhousie University, Canada
Bogumił Kamiński	Warsaw School of Economics, Poland
Christer Carlsson	Abo Akademi, Finland
Christof Weinhardt	Universität Karlsruhe, Germany
Colin Eden	Strathclyde Business School, UK
Daniel Druckman	George Mason University, USA
Douglas Vogel	Harbin Institute of Technology, China
Etiënne Rouwette	Radboud University, The Netherlands
Ewa Roszkowska	University of Białystok, Poland
Fran Ackermann	Curtin Business School, Australia
Francisco Chiclana	De Montfort University, UK
Fuad Aleskerov	National Research University HSE, Russia
Gert-Jan de Vreede	University of South Florida, USA
Ginger Ke	Memorial University of Newfoundland, Canada
Gregory Kersten	Concordia University, Canada
Guy Olivier Faure	University of Paris V–Sorbonne, France
Hannu Nurmi	University of Turku, Finland
João Clímaco	University of Coimbra, Portugal
John Zeleznikow	Victoria University, Australia
José Maria Moreno-Jiménez	Zaragoza University, Spain
Katia Sycara	Carnegie Mellon University, USA
Keith Hipel	University of Waterloo, Canada

Contents

Preference Modeling for Group Decision and Negotiations

Solving Multicriteria Group Decision-Making (MCGDM) Problems Based on Ranking with Partial Information

Eduarda Asfora Frej$^{(\boxtimes)}$ ⓘ, Adiel Teixeira de Almeida ⓘ,
and Lucia Reis Peixoto Roselli

CDSID - Center for Decision Systems and Information Development,
Universidade Federal de Pernambuco – UFPE, Av. Acadêmico Hélio Ramos,
s/n – Cidade Universitária, Recife, PE 50740-530, Brazil
eafrej@cdsid.org.br

Abstract. This paper presents an interactive Decision Support System for solving multicriteria group decision-making (MCGDM) problems, based on partial information obtained from the decision makers (DMs). The decision support tool was built based on the concept of flexible elicitation of the FITradeoff method, with graphical visualization features and a user-friendly interface. The decision model is based on searching for dominance relations between alternatives, according to the preferential information obtained from the decision-makers from tradeoff questions. A partial (or complete) ranking of the alternatives is built based on these dominance relations, which are obtained from linear programming models. The system shows, at each interaction, an overview of the process, with the partial results for all decision-makers. The visualization of the individual rankings by all DMs can help them to achieve an agreement during the process, since they will be able to see how their preferred alternatives are in the ranking of the other DMs. The applicability of the system is illustrated here with a problem for selecting a package to improve safety of oil tankers.

Keywords: Multicriteria group decision-making (MCGDM) · FITradeoff · Partial information · Ranking

1 Introduction

Preference elicitation is one of the most challenging issues related to multicriteria decision situations within the scope of additive aggregation models, due to the amount and difficulty of information generally required in the elicitation process, associated with the natural limited cognitive capacity of decision makers (DMs). According to the Multiattribute Value Theory (MAVT - [13]), alternatives are scored straightforwardly according to an additive aggregation function, as Eq. (1) shows. In (1), $v(a_j)$ is the global value of alternative a_j, k_i is the scale constant of criterion i, and $v_i(x_{ij})$ is the value function of the outcome x_{ij}, which represents the payoff of alternative a_j in criterion c_i, and n is the number of criteria.

© Springer Nature Switzerland AG 2019
D. C. Morais et al. (Eds.): GDN 2019, LNBIP 351, pp. 3–16, 2019.
https://doi.org/10.1007/978-3-030-21711-2_1

$$v(a_j) = \sum_{i=1}^{n} k_i v_i(x_{ij}) \tag{1}$$

In (1), the values of the scale constants of the criteria do not only represent the level of importance of the criteria, but involve a scaling factor as well. Due to this reason, the values of k_i should be determined based on the range of consequence values of each criterion, by taking into account the tradeoffs amongst them, and not considering only relative importance, as in outranking methods [6]. Traditional methods proposed to elicit values of criteria scaling constants such as the tradeoff elicitation procedure [13] and the swing procedure [11] require high cognitively demanding information from the DMs, and sometimes they may not be able to face such a tedious and time consuming process or to provide the preferential information in the detailed way required [3, 14, 15, 23]. Moreover, when multiple actors are involved in the decision process, it becomes even more complex, since each one seeks to exert their own influence on the process, according to their individual interests. De Almeida and Wachowicz [8] highlight the fact that decisions involving multiple DMs are much more challenging compared to decisions with only one DM, since the different viewpoints, preferences and aspirations of the multiple actors increase the complexity of the problem, in addition to the conflicting objectives already considered.

Motivated by these issues, elicitation methods that require partial/imprecise/ incomplete information about the DMs' preferences were developed. These methods consider, in general, that DMs do not have a precisely defined preference structure, and are not able to provide exact values for parameters and/or consequences values [29]. In general, the information provided can be in the form of ranking, bounds, holistic judgments and/or arbitrarily linear inequalities involving the values of criteria scaling constants or the value function. In order to compute a recommendation based on this information, the following possible synthesis steps can be applied: linear programming problems models [1, 2, 7, 15–17, 19–24]; decision rules [9, 10, 23–26]; surrogate weights [4, 5, 11, 27] and simulation and sensitivity analysis [18, 22, 23].

Most of these previous approaches lack a structured elicitation process for conducting the preference assessment together with the decision-maker, by assuming that the information in forms of ranking/bounds/linear inequalities are previously given. In this context, this works presents an interactive decision support system which aims to aid multiple criteria decision processes in which multiple decision-makers are involved, based on the Flexible and Interactive Tradeoff elicitation method, developed by de Almeida et al. [17] for individual decision making concerning choice problems. The preference elicitation of the group of DMs is conducted through a structured elicitation process based on tradeoff questions between adjacent criteria put to the DMs. Based on the answers provided by the DMs, it is possible to construct linear inequalities, which act as constraints for a linear programming problem model that searches for dominance relations between alternatives. Based on these dominance relations, a ranking of the alternatives is built for each DM [12]. With the individual alternatives' ranking of each DM, the agreement achievement process for the choice group decision problem can be aided by an analyst, which shows the possible resulting scenarios for the actors involved. This whole process is aided by graphical visualization features provided by the DSS.

The main contribution of this paper in terms of novelty is a practical decision support tool for solving multiple criteria group decision-making problems, which was developed by the authors based on the mathematical model of the FITradeoff method. The software conducts the elicitation process based on an interactive process with the DMs, with the possibility of visualizing partial results through graphics in the middle of the process. Flexibility features of the FITradeoff for individual decision-making were incorporated into the group decision system, but also other features were added to the system in order to aid collective decision-making processes. The support tool developed here is able to be used in the most different contexts and problems that involve multiple objectives and more than one decision-maker. The software is available by request to the authors.

This paper is organized as follows. Section 2 describes the partial information method for ranking alternatives used by the group decision support tool proposed here. Section 3 presents an overview of the group decision process based on the flexible and interactive tradeoff elicitation. The applicability of the system is illustrated in Sect. 4 with a problem previously approached by [28], in which a group of DMs aim to choose new measures to improve the safety of oil tankers, and thus help to prevent pollution of the sea from the ships. Finally, the main conclusions and final remarks are presented in Sect. 5.

2 Ranking Alternatives with Partial Information

Most of the partial information methods already developed in the literature try to find the best solution for a choice problem, according to the information given by the DM. When it comes to the ranking problematic, however, the complexity of the problem increases a lot, because sometimes the partial information given the DMs is not enough to rank a set of alternatives, even though a choice problem may be solved.

In this context, Frej et al. [12] developed a model to rank alternatives with partial information obtained by asking tradeoff questions to the DM. These questions are based on strict preference statements between different levels of the diverse criteria of the problem. Based on these statements, inequalities related to the criteria scale constants can be obtained, and these inequalities act as constraints for the linear programming problem (LPP) model that tries to find pairwise dominance relations between alternatives, and then build a ranking of them based on this information. The LPP model is shown below (Eqs. 2–7).

$$Max \; d_{jz} = \sum_{i=1}^{n} k_i v_i(x_{ij}) - \sum_{i=1}^{n} k_i v_i(x_{iz}), j = 1, \ldots, m; z = 1, \ldots, m; j \neq z. \tag{2}$$

$$s.t. :$$

$$k_1 > k_2 > \ldots > k_n \tag{3}$$

$$k_i v_i(x_i') \geq k_{i+1}, i = 1, \ldots, n - 1. \tag{4}$$

$$k_i v_i(x_i'') \leq k_{i+1}, i = 1, \ldots, n - 1. \tag{5}$$

$$\sum_{i=1}^{n} k_i = 1 \tag{6}$$

$$k_i \geq 0, i = 1, \ldots, n. \tag{7}$$

The objective function (2) tries to maximize the difference between the global value (calculated by Eq. (1)) of two alternatives, a_j and a_z. The decision variables are the criteria scale constants k_i, and m is the number of alternatives of the problem. The first constraint (3) is the ranking of criteria weights given by a DM, which is the first form of partial information gathered by him/her. The second and third constraints (4, 5) are related to the preference statements given by the DM obtained by the tradeoff questions made for him/her. x_i' and x_i'' are the values of criterion i for which the preferences between consequences were stated (for details, see [12]). The constraint in (6) guarantees the normalization of weights, and the constraint in (7) is regarding the non-negativity of these values.

This LPP model runs for each pair of alternative a_j, a_z in order to find dominance relations between them. Let d_{jz}^* be the optimal solution of LPP (2–7), with the maximum value of the difference between alternatives a_j and a_z. It was proved by Frej et al. [12] that if $d_{jz}^* < 0$, then a_z dominates a_j; if $d_{jz}^* \leq \varepsilon$ and $d_{zj}^* \leq \varepsilon$ (where ε is small value defined by the DM, which acts as an indifference threshold), then alternatives a_j and a_z are indifferent to each other; and if $d_{jz}^* > \varepsilon$ and $d_{zj}^* > \varepsilon$, alternatives a_j and a_z are incomparable to each other according to the current level of information obtained from the DM. Based on these dominance relations, it is possible to build a ranking of the alternatives, which may be partial if there is some incomparability arises, or complete otherwise.

For group decision-making, the information of the individual rankings (either partial or complete) during the process can help the group to achieve an agreement. Someone can see, for instance, that his/her second ranked alternative is the first place for the others, while his/her first alternative is, for instance, in the last position for the others. In this sense, sometimes the other DMs may try to convince this person to abdicate of his/her first alternative and consider the second one, since the final outcome would be better for the group as a whole. Based on this motivation, a group decision support tool based on the ranking decision model described above is presented in the next Section, in order to show how this feature can aid the group decision process, and help to find the best compromise solution.

3 Flexible and Interactive Tradeoff Elicitation for Group Decision

The group decision support tool presented here is a flexible system for aiding multi-criteria group decision-making processes, based on the ideas of the Flexible and Interactive Tradeoff support tool [7], which was originally developed for solving individual choice problems, based on the concept of potentially optimal alternatives throughout a flexible decision process, with graphical visualization features.

The decision support system allows that multiple decision-makers state their preferences at the same time, and the results of all of them can be visualized jointly. The input data of the system are the consequences matrix and the individual criteria order for each DM. Since the performance of each alternative with respect to each criterion is factual – and not preferential – information of the problem, the consequences matrix is considered to be the same for all DMs. However, the criteria order may be different for each DM, since this is indeed preferential information. Moreover, FITradeoff allows different DMs to consider different criteria in their evaluation: the weight of certain criterion which a DM does not want to consider is forced to be zero in the mathematical model, and thus this DM does not have to include this criterion in his ranking of criteria weights neither has to answer tradeoff questions concerning this criterion. This issue is illustrated later in Sect. 4.

The elicitation is conducted based on an interactive question-answering process with the decision makers. The questions are related to comparison between fictitious alternatives, with respect to which the DMs should state their preferences, by considering tradeoffs between adjacent criteria (according to the criteria order previously defined by each DM). More details of how these questions are designed in the FITradeoff system can be found in [7]. In order to give an example, let us consider a hypothetical alternative A, with an intermediate outcome x_1 in the first-ranked criterion, and the worst outcome for the others (the value function for the worst outcome is set to 0, and it is set to 1 for the best outcome, since the value function is normalized in a 0–1 scale), so the global value of A, according to Eq. (1), is $k_1 v_1(x_1)$; and now let us consider hypothetical alternative B, with the best outcome for the second-ranked criterion and the worst outcome for the others, so that the global value of B is k_2, according to Eq. (1). Depending of the value of x_1 and according to the DM's preferences, he/she will choose which one of these hypothetical alternatives is preferred to him. If A is preferred to B for a certain value of $x_1 = x_1'$, for instance, then the global value of A is greater than the global value of B, so that an inequality similar to (4) is obtained. If B is preferred to A for a certain value of $x_1 = x_1''$, then an inequality similar to (5) is obtained. Thus, these inequalities act as constraints for the LPP model (2–7), so that pairwise dominance relations are computed and a (partial or complete) ranking for the respective DM is constructed, according to level of partial information obtained from him/her until that point. At each interaction cycle, a new question is answered by this DM, so that a new inequality is obtained and incorporated to the LPP model, which runs again, so that an updated ranking of the alternatives is obtained. Therefore, at each step, with new information provided by each DM, their rankings are refined, and each DM keeps answering elicitation questions until a complete ranking of the alternatives is achieved for him/her. However, the flexibility of the process allows the DMs to visualize these partial rankings obtained during the process at any time, and this may help them to achieve an agreement before the end of the elicitation process.

The decision makers should choose in which way they want to conduct this interactive question-answering process: independently or simultaneously. In the independent elicitation process, each DM answers the elicitation questions in separate moments; the first decision maker starts the elicitation process and answers preference questions until a complete order of the alternatives is achieved for him/her or until he/she is satisfied with the results obtained. Then the second decision maker starts the elicitation process and answers preference questions until a complete order of the alternatives is achieved for him/her or until he/she is satisfied with the results obtained, and so the process goes on until all DMs had finished. In the simultaneous elicitation process, the first DM answers the first preference question, then the second DM answers his first preference question, and so on; the first DM answers the second preference question only when all DMs had answered their first preference questions. The difference between these two ways of conducting the elicitation is in the process itself; in fact, there is no difference in the results obtained by one way or another, but the simultaneous elicitation has the advantage that the partial results of all DMs can be analyzed together, since they are always in the same step of the process.

The next Section illustrates the system, based on an application of package selection for improving safety oil tankers.

4 Selecting a Package for Improving Safety Oil Tankers with FITradeoff System

The decision problem addressed here to illustrate the use of the flexible and interactive elicitation for group decision-making problems concerns the situation presented by Ulvila and Snider [28]. The authors present a negotiation situation in the International Conference on Tanker Safety and Pollution Prevention, which aimed to adopt new measures to improve the safety of oil tankers, and thus help to prevent pollution of the seas from ships. The United States and several other countries were parts of the negotiation process, and four packages of actions (U.S proposal, Package 1, Package 2 and MARPOL 73) were evaluated with respect to eleven criteria: world oil outflow (WO), oil in own waters (OOW), safety (SF), cost (CT), dollars per ton (DPT), ease of passing cost to consumer (EPC), charter party (CP), tanker surplus (TS), shipyards (SP), competitive advantage (CA), and enforceability (EF). The scores of each alternative in each criterion were defined in a 0–100 scale, for standardization purposes. The value of 0 represents the worst outcome for the criteria, and the value of 100 corresponds to the best outcome. Table 1 shows the consequences matrix for this problem.

Table 1. Consequences matrix for the tanker safety problem [28]

Packages	WO	OOW	SF	CT	DPT	EPC	CP	TS	SP	CA	EF
MARPOL 73	0	0	0	100	0	100	100	0	0	0	0
US Proposal	81	100	100	0	0	0	0	100	100	100	100
Package 2	100	0	50	95	100	95	95	0	20	20	10
Package 1	80	100	85	30	32	30	10	95	90	90	90

A total of twenty-one countries would participate in the conference, but the authors chose ten of them to illustrate the analysis. The identities of these countries – except for the US - were omitted in the description of the application, represented by A, B, C, etc. Criteria weights were elicited from a small group that represented each country in the evaluation. Table 2 shows the weights assigned for each criterion, by country. The problem was solved by Ulvila and Snider [28] based on concepts of MAVT, through an additive aggregation function, and a score for each alternative was calculated for each DM.

Table 2. Criteria weights for each country [28]

Countries	WO	OOW	SF	CT	DPT	EPC	CP	TS	SP	CA	EF
US	0.15	0.24	0.24	0.08	0.1	0	0	0	0.02	0	0.17
A	0.03	0.09	0.09	0	0.09	0	0.03	0	0.14	0.26	0.29
B	0.03	0.17	0.05	0.03	0	0	0.12	0.25	0.15	0.1	0.1
C	0.03	0.17	0.05	0.03	0	0	0.12	0.25	0.15	0.1	0.1
D	0	0.04	0.04	0.43	0.17	0.13	0.04	0	0.02	0.07	0.04
E	0.02	0.04	0.06	0.22	0.07	0.22	0.07	0.04	0.18	0.07	0
F	0	0.3	0.15	0.3	0.03	0.06	0	0	0	0	0.15
G	0.17	0.17	0.2	0.29	0.03	0	0	0	0	0	0.14
H	0	0	0	0.3	0.03	0	0.2	0.1	0	0.03	0.33
I	0.07	0.07	0.1	0.15	0.12	0	0.12	0	0	0.12	0.24

4.1 Decision Process with FITradeoff System

The first step is the input of the data of the problem. The input file should contain the consequences matrix and the criteria order for each decision-maker. The order of weights for each DM (see Table 3) was obtained based on the weights in Table 2. These weights were also used to simulate the answers of the questions made by the system for each DM. It should be highlighted here that the FITradeoff system allows that different decision-makers consider different criteria in their evaluation. The tanker safety problem addressed here illustrates such case, since in Table 2 it can be seen that some criteria weights are equal to zero for some decision makers, which means that they do not want to consider these criteria in their evaluation. For instance, the criteria EPC and TS are not considered in the evaluation of DM "A" (countries are here treated as decision-makers, for simplification purposes).

In order to input the data, the user uploads an Excel spreadsheet in the DSS, that should contain the consequences matrix – as in Table 1 – and the criteria order for each DM, as Table 3 shows. The criteria order of Table 3 was obtained based on the original values of weights shown in Table 2. Some criteria may also be tied – when they have the same weight value, and thus they receive the same index in Table 3 (see, for instance, criteria OOW and SF for DM "US").

Table 3. Criteria order for each DM

Criteria order	WO	OOW	SF	CT	DPT	EPC	CP	TS	SP	CA	EF
US	4	1	1	6	5	0	0	0	7	0	3
A	7	4	4	0	4	0	7	9	3	2	1
B	8	2	7	8	0	0	4	1	3	5	5
C	7	2	5	4	7	0	0	5	0	3	1
D	0	5	5	1	2	3	5	0	9	4	5
E	10	8	7	1	4	1	4	8	3	4	0
F	0	1	3	1	6	5	0	0	0	0	3
G	3	3	2	1	6	0	0	0	0	0	5
H	0	0	0	2	5	0	3	4	0	5	1
I	7	7	6	2	3	0	3	0	0	3	1

Figure 1 shows an example of the first elicitation question made for the first DM in the order of the input file, which is "US" (see the box "Current Decision Maker" in the top right side of Fig. 1). The elicitation questions ask the DM to choose between two consequences, A and B, by considering tradeoffs amongst different criteria. In the case of Fig. 1, consequence A has an intermediate outcome for criterion C1 (OOW) and the worst outcome for the other criteria; and consequence B has the best possible outcome for criterion C7 (SP) and the worst outcome for all other criteria. At this point, the DM can choose one of the options given to answer the question: preference for consequence A, preference for consequence B, indifference between the two consequences, or even "no answer", if he/she does not want to answer that question, maybe because it is too difficult for him/her, or simply because the DM thinks he is not able to answer the question in a consistent way. In this case, another question will be formulated, without

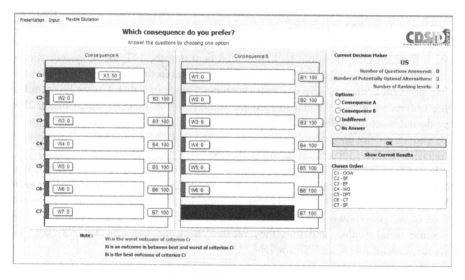

Fig. 1. Elicitation question

loss of information in the process. The box "Current Decision Maker" in the right side of the screen shows some information regarding the elicitation process of the current DM: number of questions answered, which is zero in this case, since this is the first questions made for him; number of potentially optimal alternatives (2) at that point; and number of ranking levels (3).

By clicking on the button "Show Current Results", in the right side of the screen in Fig. 1, it is possible to visualize the partial results obtained for each DM until that point. Figure 2 shows the screen of FITradeoff system with the partial results, with two tables: the first one shows the potentially optimal alternatives for each DM, calculated as in the standard FITradeoff method [7], and the second one shows the ranking of the alternatives, which can be a partial ranking or a complete ranking, obtained by the dominance relations found by the LPP model presented in Sect. 2.

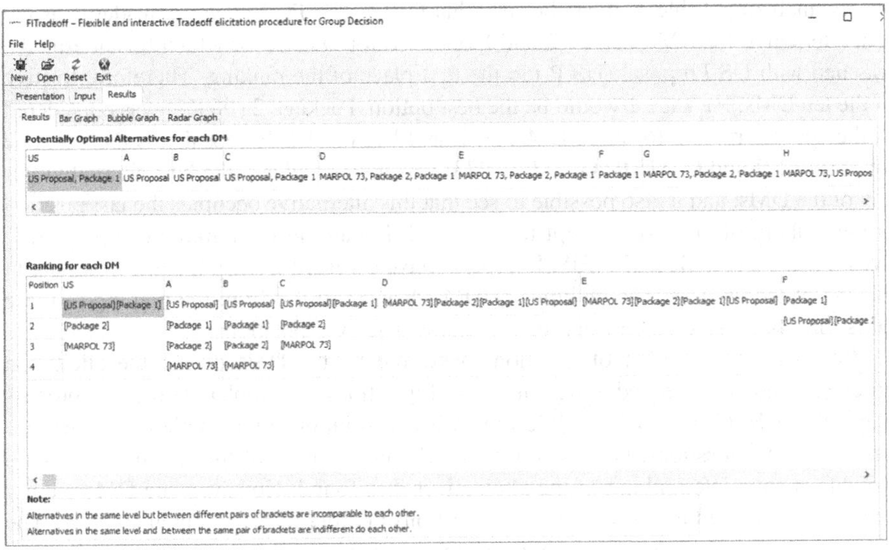

Fig. 2. Partial results

By analyzing Fig. 2, it is possible to see that, at this point of the process, a complete order of the alternatives has been achieved for some of the DMs, such as "A" and "B", just with the information of the ranking of criteria weights. For some other DMs, a partial order of the alternatives was reached at this point (see the rankings for DMs "US", "C" and "F"). For other DMs, no ranking levels were established at this step.

The idea in FITradeoff elicitation process is that as the DMs provide more preferential information, the weight space is updated in such a way that the ranking becomes more complete at each interaction. By assuming that the decision makers would follow the standard decision process, by answering to the questions made in

FITradeoff according to the weights given in Table 2, the results achieved for each DM would be those displayed in Table 4.

Table 4. Final results for each DM

DM		US	A	B	C	D	E	F	G	H	I
Interaction cycles		14	0	0	16	14	29	4	12	16	10
Final ranking	1	PCK1	US P.	US P.	US P., PCK1	PCK2	PCK2	PCK1	PCK1	PCK2	PCK1
	2	US P.	PCK1	PCK1	PCK2	MP73	PCK1	US P.	US P.	PCK1	US P.
	3	PCK2	PCK2	PCK2	MP73	PCK1	US P.	PCK2	PCK2	MP73	PCK2
	4	MP73	MP73	MP73	–	US P.	MP73	MP73	MP73	US P.	MP73

By analyzing Table 4, it can be seen that Package 1 (PCK1) was considered as the best alternative for DMs US, F, G and I, and also for DM C, for which this alternative was tied with US Proposal (US P.) in the first place of the ranking. Therefore, for five of the ten DMs, Package 1 would be the best option. Package 2 (PCK2) was considered as the best alternative for three DMs, D, E and H. And US Proposal was the best one for DMs A, B and C. MARPOL 73 (MP73) was not considered the best alternative for any of the DMs, and it also possible to see that this alternative occupies the last position in the ranking of all DMs, except for D and H. Regarding the ranking of the alternatives, it can be seen that DMs US, F, G and I have obtained exactly the same ranking in the end of the process. The rankings of DMs A and B differ from the ranking of these DMs only because of the switch of the first and second positions.

Regarding the number of questions answered by the DMs during the elicitation process, some DMs needed a high number of questions to complete the whole process (e.g. DMs US, C, E and H), while others had a solution found without the need to answer to any questions, because only with the information of the ranking of criteria weights, a complete ranking of the alternatives was found. However, the DMs are not necessarily required to answer all these questions in order to reach a final solution. The flexibility features of the system allow them to maybe reach an agreement before the end of the process, as discussed in the next topic.

4.2   Discussion of Results

The key feature of the FITradeoff system is its flexibility. DMs can visualize partial results at each step during the process, and they can possibly reach an agreement before the end of the elicitation process, based on the analysis of the partial results provided by the system in tables and graphical tools.

For instance, let us consider the partial results obtained just with the information of ranking of criteria weights, i.e., before the question-answering process starts. The results of POA (Potentially Optimal Alternatives) and ranking at this step were shown in Fig. 2. The system provides individual graphical visualization of the best alternatives for each DM, and also a collective graphic with the best alternatives for all DMs, as shown in Fig. 3.

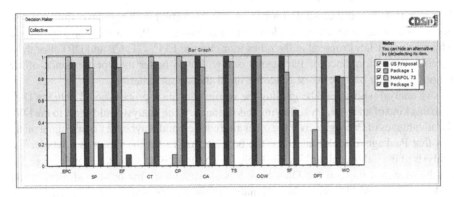

Fig. 3. Collective bar graphic for all DMs

At this point, all the four alternatives of the problem are presented in the collective graphic, since they are potentially optimal for at least one DM. By analyzing Fig. 3, it is possible to see that alternative MARPOL 73 has the best performance for three criteria (EPC, CT and CP), but, on the other hand, it has the worst possible outcome for the eight other criteria. It should be noticed that when an alternative does not appear in the graphic for certain criterion, it means that this alternative has the worst outcome for this criterion; e.g. for criterion TS, alternatives MARPOL 73 and Package 2 do not appear in the graphic, because they have the worst possible outcome in these criterion (as can be seen in Table 1). Another feature of the flexibility of the DSS is the possibility of deselect alternatives in the box in the right side of the screen, so that the graphic shows only a subset of alternatives, chosen by the users. Since MARPOL 73 has a very bad outcome in most of the criteria, one could deselect this alternative in order to analyze just the other three alternatives, as Fig. 4 shows.

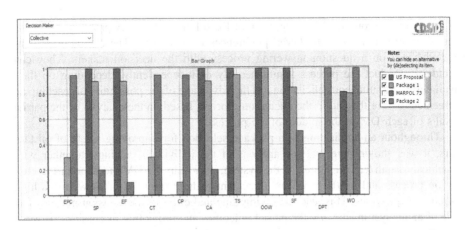

Fig. 4. Collective bar graphic for all DMs, without MARPOL 73

Figure 4 illustrates the comparison of the three better alternatives for this problem. It can be seen that Package 2 has a good performance in five of the eleven criteria, and has a very low outcome in the six other criteria. Alternatives US Proposal and Package

1 appear to have very close performances for this problem. They have similar – and good – outcomes in seven of the eleven criteria. However, Package 1 seems to have an advantage over US Proposal in the other four criteria: EPC, CT, CP and DPT. For these criteria, the performance of Package 1 is low, but is better than performance of US Proposal, which has the worst outcome in all these four criteria.

An analyst who should be guiding the whole elicitation process could help the DMs with this kind of analysis. By following this rationality, the analyst could show to the DMs the advantages of Package 1 over the other alternatives, and maybe they could agree on the fact that Package 1 could indeed be the best alternative for the group as a whole. The analysis of the individual rankings in Table 4 also supports an advantage for Package 1. It can be seen that most of the DMs have this alternative in first or second place of the ranking. Only DM "D" ranks this alternative in third place, and none of the DMs ranks Package 1 in the last position of the ranking. It should be highlighted here that this analysis is available since the beginning of the decision making process, when the DMs have answered no questions yet, which shows that the elicitation could be shortened based on graphical analysis. Behavioral studies on graphical visualization, using neuroscience tools, have been conducted [30] in order to improve the design and insightful information for the interaction between analyst and DMs. These studies are being applied in order to facilitate the process for reaching agreement between DMs [31].

It may happen, however, that the individuals do not reach an agreement even with the support of the analyst, especially in cases in which extreme opinions arise; in this case, some aggregating decision rules (such as *minimax, maximin, minimax regret*, etc.) could be used in order to find the best collective solution for the group as a whole. As for future studies, these rules should be better analyzed for implementation in the decision support tool.

5 Conclusions

In this work, a group decision support tool for solving MCDM/A problems based on partial information about the DMs' preferences is developed. The elicitation is conducted based on a question-answering process with the decision makers. They can decide to conduct the process simultaneously or independently, depending on their availability. Simultaneous elicitation has the advantage of the possibility of reaching an agreement before the end of the elicitation process, based on the analysis of the partial results of each DM in the middle of the process.

Throughout an application of a package selection for improving safety of oil tankers, it was shown here that the analysis of the individual rankings, together with graphical visualization features of the system, may help the DMs to shorten the elicitation process and maybe reach an agreement without the need of answering a high number of questions. This leads to saving time and effort of the decision makers. The role of an analyst in showing the partial outcomes and promoting discussions between the DMs is also crucial in this process.

For future research, it would be interesting to investigate the validation of the FITradeoff method through simulation studies, in order to support the empirical results obtained by applications already made, and also to measure the actual saving in terms of time and effort of the DMs.

Acknowledgments. The authors are most grateful for CNPq and CAPES, for the financial support provided.

References

1. Ahn, B.S., Park, K.S.: Comparing methods for multiattribute decision making with ordinal weights. Comput. Oper. Res. **35**, 1660–1670 (2008). https://doi.org/10.1016/j.cor.2006.09.026
2. Athanassopoulos, A.D., Podinovski, V.V.: Dominance and potential optimality in multiple criteria decision analysis with imprecise information. J. Oper. Res. Soc. **48**, 142–150 (1997). https://doi.org/10.1057/palgrave.jors.2600345
3. Belton, V., Stewart, T.: Multiple Criteria Decision Analysis: An Integrated Approach. Springer, Heidelberg (2002). https://doi.org/10.1007/978-1-4615-1495-4
4. Danielson, M., Ekenberg, L.: A robustness study of state-of-the-art surrogate weights for MCDM. Group Decis. Negot. **26**, 677–691 (2017). https://doi.org/10.1007/s10726-016-9494-6
5. Danielson, M., Ekenberg, L., Larsson, A., Riabacke, M.: Weighting under ambiguous preferences and imprecise differences in a cardinal rank ordering process. Int. J. Comput. Intell. Syst. **7**, 105–112 (2014). https://doi.org/10.1080/18756891.2014.853954
6. de Almeida, A.T., Cavalcante, C.A.V., Alencar, M.H., Ferreira, R.J.P., Almeida-Filho, A.T., Garcez, T.V.: Multicriteria and multiobjective models for risk, reliability and maintenance decision analysis. International Series in Operations Research & Management Science, vol. 231. Springer, New York (2015). https://doi.org/10.1007/978-3-319-17969-8
7. de Almeida, A.T., de Almeida, J.A., Costa, A.P.C.S., de Almeida-Filho, A.T.: A new method for elicitation of criteria weights in additive models: flexible and interactive tradeoff. Eur. J. Oper. Res. **250**, 179–191 (2016). https://doi.org/10.1016/j.ejor.2015.08.058
8. de Almeida, A.T., Wachowicz, T.: Preference analysis and decision support in negotiations and group decisions. Group Decis. Negot. **26**, 649–652 (2017). https://doi.org/10.1007/s10726-017-9538-6
9. Dias, L.C., Clímaco, J.N.: Dealing with imprecise information in group multicriteria decisions: a methodology and a GDSS architecture. Eur. J. Oper. Res. **160**, 291–307 (2005). https://doi.org/10.1016/j.ejor.2003.09.002
10. Dias, L.C., Clímaco, J.N.: Additive aggregation with variable interdependent parameters: the VIP analysis software. J. Oper. Res. Soc. **51**, 1070–1082 (2000). https://doi.org/10.1057/palgrave.jors.2601012
11. Edwards, W., Barron, F.H.: SMARTS and SMARTER: improved simple methods for multiattribute utility measurement. Organ. Behav. Hum. Decis. Process. **60**, 306–325 (1994). https://doi.org/10.1006/obhd.1994.1087
12. Frej, E.A., de Almeida, A.T., Cabral, A.P.C.S.: Using data visualization for ranking alternatives with partial information and interactive tradeoff elicitation. Oper. Res. (2019). https://doi.org/10.1007/s12351-018-00444-2
13. Keeney, R.L., Raiffa, H.: Decision Analysis with Multiple Conflicting Objectives. Wiley, New York (1976)
14. Kirkwood, C.W., Corner, J.L.: The effectiveness of partial information about attribute weights for ranking alternatives in multiattribute decision making. Organ. Behav. Hum. Decis. Process. **54**, 456–476 (1993). https://doi.org/10.1006/obhd.1993.1019
15. Kirkwood, C.W., Sarin, R.K.: Ranking with partial information: a method and an application. Oper. Res. **33**, 38–48 (1985). https://doi.org/10.1287/opre.33.1.38
16. Malakooti, B.: Ranking and screening multiple criteria alternatives with partial information and use of ordinal and cardinal strength of preferences. IEEE Trans. Syst. Man Cybern. Part A: Syst. Hum. **30**, 355–368 (2000). https://doi.org/10.1109/3468.844359

17. Mármol, A.M., Puerto, J., Fernández, F.R.: Sequential incorporation of imprecise information in multiple criteria decision processes. Eur. J. Oper. Res. **137**, 123–133 (2002). https://doi.org/10.1016/s0377-2217(01)00082-0

18. Montiel, L.V., Bickel, J.E.: A generalized sampling approach for multilinear utility functions given partial preference information. Decis. Anal. **11**, 147–170 (2004). https://doi.org/10.1287/deca.2014.0296

19. Mustajoki, J., Hämäläinen, R.P., Salo, A.: Decision support by interval SMART/SWING - incorporating imprecision in the SMART and SWING methods. Decis. Sci. **36**, 317–339 (2005). https://doi.org/10.1111/j.1540-5414.2005.00075.x

20. Park, K.S.: Mathematical programming models for characterizing dominance and potential optimality when multicriteria alternative values and weights are simultaneously incomplete. IEEE Trans. Syst. Man Cybern. Part A: Syst. Hum. **34**, 601–614 (2004). https://doi.org/10.1109/tsmca.2004.832828

21. Park, K.S., Kim, S.H.: Tools for interactive multiattribute decision-making with incompletely identified information. Eur. J. Oper. Res. **98**, 111–123 (1997). https://doi.org/10.1016/0377-2217(95)00121-2

22. Salo, A.A., Hämäläinen, R.P.: Preference assessment by imprecise ratio statements. Oper. Res. **40**, 1053–1061 (1992). https://doi.org/10.1287/opre.40.6.1053

23. Salo, A.A., Hämälainen, R.P.: Preference ratios in multiattribute evaluation (PRIME)-elicitation and decision procedures under incomplete information. IEEE Trans. Syst. Man Cybern. Part A: Syst. Hum. **31**, 533–545 (2001). https://doi.org/10.1109/3468.983411

24. Salo, A.A., Punkka, A.: Rank inclusion in criteria hierarchies. Eur. J. Oper. Res. **163**, 338–356 (2005). https://doi.org/10.1016/j.ejor.2003.10.014

25. Sarabando, P., Dias, L.C.: Simple procedures of choice in multicriteria problems without precise information about the alternatives' values. Comput. Oper. Res. **37**, 2239–2247 (2010). https://doi.org/10.1016/j.cor.2010.03.014

26. Sarabando, P., Dias, L.C.: Multiattribute choice with ordinal information: a comparison of different decision rules. IEEE Trans. Syst. Man Cybern. Part A: Syst. Hum. **39**, 545–554 (2009). https://doi.org/10.1109/tsmca.2009.2014555

27. Stillwell, W.G., Seaver, D.A., Edwards, W.: A comparison of weight approximation techniques in multiattribute utility decision making. Organ. Behav. Hum. Perform. **28**, 62–77 (1981). https://doi.org/10.1016/0030-5073(81)90015-5

28. Ulvila, J.W., Snider, W.D.: Negotiation of international oil tanker standards: an application of multiattribute value theory. Oper. Res. **28**, 81–96 (1980). https://doi.org/10.1287/opre.28.1.81

29. Weber, M.: Decision making with incomplete information. Eur. J. Oper. Res. **28**, 44–57 (1987). https://doi.org/10.1016/0377-2217(87)90168-8

30. Roselli, L.R.P., de Almeida, A.T., Frej, E.A.: Decision neuroscience for improving data visualization of decision support in the FITradeoff method. Oper. Res. (2019). https://doi.org/10.1007/s12351-018-00445-1

31. Roselli, L.R.P., Frej, E.A., de Almeida, A.T.: Neuroscience experiment for graphical visualization in the FITradeoff Decision Support System. In: Chen, Y., Kersten, G., Vetschera, R., Xu, H. (eds.) GDN 2018. LNBIP, vol. 315, pp. 56–69. Springer, Cham (2018). https://doi.org/10.1007/978-3-319-92874-6_5

How to Address Multiple Sources
of Influence in Group Decision-Making?
From a Nonordering to an Ordering Approach

Hang Luo$^{(\boxtimes)}$ (ID)

Peking University, Beijing 100871, China
`hang.luo@pku.edu.cn`

Abstract. We consider settings of group decisions where agents'
choices/preferences are influenced (and thus changed) by each other.
Previous work discussed at length the influence among agents and how
to obtain the result of influence resulting choice/preference but mainly
considered the influence of one agent at a time or the simultaneous influence of more than one agent but in a nonordering (such as unidimensional
utility or belief, binary opinion or choice) context. However, the question
regarding how to address multiple influences in an ordering (specifically,
ordinal preference) context, particularly with varied strengths (stronger
or weaker) and opposite polarities (positive or negative), remains. In
this paper, we extend classical *social choice functions*, such as the Borda
count and the Condorcet method, to signed and weighted *social influence functions*. More importantly, we extend the KSB (Kemeny) distance metric to a *matrix influence function*. Firstly, we define the rule
for transforming each preference ordering into a corresponding matrix
(named the *ordering matrix*) and set a metric to support the computation of the distance between any two ordering matrices (namely, any
two preferences). Then, the preference (theoretically existing) that has
the smallest weighted sum of distances from all influencing agents' preferences will be the resulting preference for the influenced agent. As the
weight of influencing agents can be either positive or negative (as friends
or enemies) in a real-world situation, it will play a role in finding the
"closest" possible preference from the positively influencing preferences
and finding the "farthest" possible preference from the negatively influencing preferences.

Keywords: Group decision-making · Preference ordering ·
Social choice function · Social network · Social influence function

1 Introduction

The influence among behaviors and preferences in a multi-agent (such as group
decision-making) system is quite common and has been noted and discussed
by scholars from varied disciplines, including artificial intelligence (particularly

© Springer Nature Switzerland AG 2019
D. C. Morais et al. (Eds.): GDN 2019, LNBIP 351, pp. 17–32, 2019.
https://doi.org/10.1007/978-3-030-21711-2_2

multi-agent systems), economics, and decision theory [3–12,19,20,23–27]. In the context of a group decision, an agent always has the motivation to influence[1] others' choices or preferences to make them support his or her own preferred alternative (candidate), thus increasing the possibility that his or her preferred alternative becomes the result of the collective choice. Further, influences in real-world settings are diversified in both polarity and strength, such as a positive influence from a friend (ally) versus a negative influence from an enemy (opponent) [23,24] and a strong influence from an intimate friend (family, relative) versus a weak influence from a common friend [23]. Actually, the strength of influence can be affected by many factors, for example: the actual or perceived power, force or authority of the influencer [23]; or how much the influenced one trusts the influencer, considering that the higher the trust is, the greater is the weight of the influence [12].

Previous work discussed at length the influence among agents (decision-makers), such as the influence of one agent at a time on another agent or the simultaneous influence of more than one agent but regarding only cardinal utility, belief, opinion, or choice [3–12,19,20,23–27] rather than ordinal preference. Actually, in the above mentioned work, agents' preference orderings are not specifically presented and discussed; thus, influences among agents can only work in a nonordering form directly from the influencing agents' (cardinal) utilities, beliefs, opinions or choices to the influenced agent's utility, belief, opinion or choice but not from the influencing agents' (ordinal) preferences among all alternatives to the influenced agent's preference. Thus, the above influence models can only support the influence study in the context of a nonranked group decision (such as the plurality) but not the ranked group decision (such as the Borda count and the Condorcet method). However, it is widely recognized that non-ranked methods have considerable drawbacks compared with ranked methods, as most of the preference information will be ignored.

In addition, previous work [3–12,20,23,27] considered the variation in strength of different influences but barely considered the variation in polarity of different influences[2] (similar to assuming that all of influences are positive). In fact, positive and negative influences are both common, and together constitute the real-world influence.

Therefore, determining how to address influences of multiple agents simultaneously in an ordering (more specifically, ordinal preference) approach, especially with varied strengths (stronger or weaker) and opposite polarities (positive or negative) and to obtain the resulting preference of an influenced agent are meaningful topics that have not been yet truly discussed.

---

[1] Say in the form of convincing.

[2] [23] considered the polarity of individual influences but just in a nonordering (cardinal utility) context.

2 Background

2.1 Social Choice Theory

Social choice refers to combining individual choices, preferences, or welfare to reach a collective choice, preference or social welfare. Social choice functions include any method that can help determine the social choice based on individuals', which mainly include group decision (most typically, voting), negotiation and so on. A widely used classification of social choice functions considers ranked and nonranked methods. In nonranked methods, only one most preferred alternative (candidate) can be chosen by each individual, while ranked methods allow a full ordering among all alternatives (candidates) to be displayed by each individual.

Nonranked Social Choice Functions

Plurality is the most simple and natural solution to choose a winner: each agent casts one vote for only one alternative, and the alternative with the highest number of votes wins [16].

Majority is relatively stricter, requiring an alternative to receive more than half of the votes, not just the highest number, to be the winner [16]. If no alternative can obtain more than half of the votes, the voting process cannot be finished in just one round. There are other remedies for this situation, such as the two-round system (namely, the second ballot).

Ranked Social Choice Functions

Borda count allows each agent to rank all alternatives on his or her ballot. The alternatives each obtain a number of scores based on their ranks on all of the agents' ballots according to a score allocation scheme, where the scores decrease with respect to the rank [16]. A typical scheme is as follows: if there are m alternatives, the i-th ranked alternative is allocated a score of $m - i + 1$, for instance, scores of $m, m - 1, ..., 1$ correspond to the 1st, 2nd, ..., and mth ranked alternatives [16]. The scores each alternative receive from all agents' ballots are summed, and the alternative with the highest summed score is declared the winner.

Condorcet methods allow each agent to rank all alternatives according to his or her preference. The alternative that is pairwise preferred to all other alternatives by the majority of agents is named the *Condorcet winner*. The Condorcet method for m alternatives in essence is to hold $C_m^2 = \frac{1}{2}m(m - 1)$ majority elections between all possible pairs of alternatives [16].

2.2   Social Network Theory

A social network is a social structure made up of a set of social actors (called nodes or agents) and a set of bilateral relations between them (called ties or links). The field of social networks has emerged as a very successful interdisciplinary area of research, drawing attention from the sociology [13–15,18,19], mathematics (graph theory) [17], computer science (AI) [12,25–27], economics [3–11,20], politics (international relations) [23], and so on. According to previous work, interpersonal ties are generally varied in strength as stronger or weaker [13–15,22] and opposite in polarity, namely, positive or negative [17,18,23,24].

Weak and Strong Ties. Granovetter [15] stated, "The 'strength' of an interpersonal tie is a linear combination of the amount of time, the emotional intensity, the intimacy (or mutual confiding), and the reciprocal services which characterize each tie", and argued that weak ties are usually more important than strong ties, as strongly linked persons (such as our families and close friends) tend to stay in the same circle with us, and the information they possess usually overlaps considerably with what we already grasp. Thus, weak ties are responsible for the majority of new information transmission [15], and such an idea can be concluded as the *strength of weak ties*.

Krackhardt [22] proposed a contrary idea of the *strength of strong ties* and stated, "People resist change and are uncomfortable with uncertainty. Strong ties constitute a base of trust that can reduce resistance and provide comfort in the face of uncertainty", thus it will be argued that change is not facilitated by weak ties but rather by strong ties. Though weak ties are powerful ways to convey awareness of new things, they are weak at transmitting behaviors that are in some way risky or costly to adopt [22]. Briefly, the *weak/strong ties paradox* exists. Actually, Granovetter [13] also admitted that "weak ties provide people with access to information and resources beyond those available in their own social circle; but strong ties have greater motivation to be of assistance and are typically more easily available".

Positive and Negative Ties. There are positive ties with friends, families and so on; however, it is also possible to face negative ties in real-world settings with enemies, opponents, or any person with a negative appreciation. The most classical work is the *structural balance theory* proposed by Heider [18], which discussed the balance or unbalance of a triangular relationship composed of three bilateral relationships (ties) that can be positive or negative (respectively representing amity or enmity relations). The product of signs of negative or positive ties determines the balanced situation of a triangular relationship such that if the product is negative, then the triangule is unbalanced; otherwise, it is balanced, which is based on fundamental psychological rules, such as the psychological transitivity: "My friend's friend is my friend" and "My friend's enemy is my enemy".

Then, Harary [17] discussed the networks composed of positive/negative ties from the mathematical perspective, developed the *signed graphs*, and further discussed the conditions for a signed graph to be balanced: only if every triangular relation in this signed graph is balanced, it is called balanced; if at least one triangular relation in this signed graph is ever unbalanced, then it is unbalanced. It is proven that if a network of interrelated positive and negative ties is balanced, then it should consist of two subnetworks such that each has positive ties between all its nodes and negative ties between nodes in distinct subnetworks, which means a social system is split into two cliques[3]; however, one of the two subnetworks may be empty, which might occur in very small networks [17].

In conclusion, we should note that the social ties discussed above were assumed to be symmetric, while this may not be the case in influencing relations.

2.3   Social Influence Model

Social influence is a model capturing the influencing relations among agents when behaviors or preferences of agents are affected by others. Social influence may take many forms and can be seen in conformity, peer pressure, leadership, persuasion [4], and so on.

Recently, in the fields of artificial intelligence (particularly multi-agent system) [12,25–27], economics [3–11,19,20] and politics [23,24], the dynamics of influence have been intensely studied, especially in the framework of social networks, using ties between agents to represent influencing relationships between them. In fact, the influence is usually asymmetrical [23]: there is no reason that the way agent i influences agent j should be the same as the way agent j influences agent i (thus, directed ties but not undirected ties are needed in a graph). Further, the influence is usually a mutual and iterative process [8]: it may well be the case that agent i influences agent j, which then influences agent k, which in turn influences agent i.

Many of these models assume a unidimensional value such as belief or utility for each agent, which will be affected by his or her "neighbors", depending on the social structure. In this line, the model of "reaching a consensus" [3] and the model of "decisional power" [19] might be foundational. More recently, Jackson [20] systematically discussed a social network environment where one agent's utility will be influenced (in the form of learning) by the agent's own and all other agents' utilities according to a weight allocation. Actually, the influence among all agents in a social network can be represented as a matrix composed of entries that each represent the weight of influence between two agents.

The study on influence by Grabisch and Rusinowska [6–11] should also be mentioned. They discussed and compared influence functions, command game [9] and follower functions [10], that are embedded in social networks [6–8]; they assumed that players are to make a binary choice, where each has an inclination to say either "yes" or "no", and due to the influence of other players, the choice of a player may be different from his or her original inclination; they defined such

---

[3] The most classical example is the formation of two conflicting alliances Central Powers and Allied Powers before World War I, which has been fully discussed by [1].

a transformation as influence. Besides, they [7] generalized the yes/no model to a multi-choice framework, but each choice is assigned with a number (utility value), thus, the influencing subject and influenced object are still cardinal but not ordinal preference; they [7] also discussed both positive and negative influences but in a group (coalitional) way rather an individual way; further, they [8] generalized a yes/no model of influence in a social network with a single step of mutual influence to a framework with iterated influence.

Most recently, [12] proposed a model combining judgment aggregation and social networks, where an agent's binary (yes/no) opinions will be affected by his or her neighbors in the network depending on the trust the agent has in them; however a binary context was still assumed, not mentioning preference ordering; though they discussed individual influences, negative influences of neighbors were not discussed, rather, it was assumed that all neighbors' opinions have positive influences.

3   Matrix Influence Function: Modeling

To successfully support the consideration of a multiple-influence mechanism in an ordering context, we introduce a *matrix influence function*[4], making use of the classical KSB (Kemeny) distance metric [2, 21, 28, 29] and extending it into a signed and weighed version. In fact, [2] generalizes the work by [21] that obtains a distance measure on strict partial orderings as the unique metric satisfying several natural axioms [28]. This metric, called the KSB metric (named based on the initials of the three contributors [2, 21]), is defined in terms of a matrix representation of preference orderings.

Definition 1. *(Group Decision-making Society with Mutual-Influence on Preference) Assume a society* $\mathbb{S} = \{\mathbb{N}, \mathbb{M}, \mathbb{P}, \mathbb{W}\}$, *where* $\mathbb{N} = \{1, 2, ..., n\}$ *is the set of all agents (a general term for decision-makers, voters, game players, etc.);* $\mathbb{M} = \{o_1, o_2, ..., o_m\}$ *is the set of all alternatives (candidates);* $\mathbb{P} = \{P_{(1)}, P_{(2)}, ..., P_{(n)}\}$ *is the set of agents' preferences; all possible preference orderings according to the set of alternatives* $\mathbb{M}$ *will include* $m!$ *kinds , the set can thus be defined as* $\mathbb{P}[\mathbb{M}] = \{p_1, p_2, ..., p_{m!}\}$; $\mathbb{W}$ *is the matrix whose entries are weights of influence between each of two agents,* $\mathbb{W} = [w_{(i,j)}]$ $(i, j \in \mathbb{N})$, *in which* $w_{(i,j)}$ *means the weight of influence from agent* i *to agent* j, *the weight value indicates both the strength and polarity of the influence,* $w_{(i,j)} > 0$ *means a positive influence,* $w_{(i,j)} < 0$ *means a negative influence, and* $w_{(i,j)} = 0$ *means there is no influence from agent* i *to agent* j.

It should be noted that the weight of influence might not be symmetrical between two agents, i.e., $w_{(i,j)} \neq w_{(j,i)}$. For instance, a very common real-world scenario is that you think someone is your friend, but he or she does not think similarly but thinks of you as a bother.

---

[4] As this influence rule first asks that all feasible preference orderings be transformed into corresponding matrices, it can thus be named the *matrix influence function*.

Definition 2. *(Ordering Matrix) is a matrix transformed from preference ordering, which can be written as* $OM = [om_{o,o'}]$, *in which* o, o' ($o, o' \in M = \{o_1, o_2, ..., o_m\}$) *are two different alternatives. An ordering matrix is in canonical form if the column and row are ordered lexicographically with the set of alternatives:*

$$
OM = \begin{array}{c} \\ o_1 \\ o_2 \\ ... \\ o_m \end{array}
\begin{array}{cccc}
o_1 & o_2 & ... & o_m \\
\left(\begin{array}{cccc}
om_{o_1,o_1} & om_{o_2,o_1} & \cdots & om_{o_m,o_1} \\
om_{o_1,o_2} & om_{o_2,o_2} & \cdots & om_{o_m,o_2} \\
... & ... & ... & ... \\
om_{o_1,o_m} & om_{o_2,o_m} & \cdots & om_{o_m,o_m}
\end{array} \right)
\end{array}
$$

For the preference ordering of agent i, *namely,* $P_{(i)}$, *its corresponding ordering matrix is* $OM_{P_{(i)}} = [om_{o,o'}^{P_{(i)}}]$:

$$
om_{o,o'}^{P_{(i)}} = \begin{cases} 1 & \text{if } o \text{ is strictly preferred to } o' \text{ by } P_{(i)} \\ -1 & \text{if } o' \text{ is strictly preferred to } o \text{ by } P_{(i)} \quad i \in \mathbb{N} \\ 0 & \text{if } o \text{ is just } o' \text{ itself} \end{cases}
$$

In the above, we assume that there is a strict ordering among all different alternatives.

Definition 3. *(Distance between Preference Orderings) is a distance in the sense of the number of swapping adjacent alternatives to make two preference orderings identical. Assume there are two agents* i *and* j *with respective preferences* $P_{(i)}$ *and* $P_{(j)}$; *let* $om_{o,o'}^{P_{(i)}}$ *and* $om_{o,o'}^{P_{(j)}}$ *be their corresponding entries from respective ordering matrices* $OM_{P_{(i)}}$ *and* $OM_{P_{(j)}}$; *then, the distance between* $P_{(i)}$ *and* $P_{(j)}$ *is (let Dis be the distance function):*

$$
Dis(P_{(i)}, P_{(j)}) = \sum_{o \in M} \sum_{o' \in M} |om_{o,o'}^{P_{(i)}} - om_{o,o'}^{P_{(j)}}| \quad i,j \in \mathbb{N}
$$

Definition 4. *(Matrix Influence Rule) Let* $P'_{(i)}$ *be the preference of agent* i *after being influenced; the result is one possible preference ordering* ($p \in \mathbb{P}[M]$) *which has the minimum weighted sum of distances from all influencing preferences:*

$$
P'_{(i)} = \arg \min_{p \in \mathbb{P}[M]} [\sum_{j \in \mathbb{N}} w_{(j,i)} Dis(P_{(j)}, p)] \quad i \in \mathbb{N}
$$

If this influence function considers the iteration of influence and multiperiod interactions, let $P_{(i)}(t+1)$ *be the preference of agent* i *after the tth mutual influence, and let* $P_{(j)}(t)$ *be the preference of agent* j *at the tth mutual influence:*

$$
P_{(i)}(t+1) = \arg \min_{p \in \mathbb{P}[M]} [\sum_{j \in \mathbb{N}} w_{(j,i)} Dis(P_{(j)}(t), p)] \quad i \in \mathbb{N}
$$

As the weight of influence can be either positive or negative (such as a positive influence from a friend or a negative influence from an enemy), it will play a role in finding the "closest" possible preference compared with positively influencing agents' preferences and finding the "farthest" possible preference compared with negatively influencing agents' preferences.

4   An Example of Multiple Sources of Influence

We propose an example to display and compare how different approaches, including the nonordering approach and the ordering approach, can be used to address the simultaneous influence of more than one agent. As shown in Fig. 1, we assume there are four agents (named M, F_1, F_2, E) making choices on three alternatives (a, b, c). We start from the perspective of agent M: agent M ("Me") is collectively influenced at the same time by agents F_1 (Friend 1), F_2 (Friend 2), E (Enemy), and usually M ("Myself"), with preferences of $c \succ b \succ a$, $a \succ c \succ b$, $b \succ c \succ a$, and $a \succ b \succ c$ and with weights of influence of 4, 2, -1, and 2. Usually, a person is positively influenced by his or her friends (with friends, some might have closer relations than others, differentiated by a larger or smaller value of the positive weight), negatively influenced by his or her enemies (which means trying to be far from enemies' preferences), and positively influenced by (or referring to) his or her own former preference.

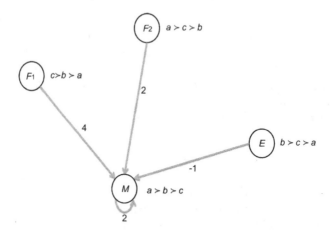

Fig. 1. An example of simultaneous influence of more than one agent

It is quite common that we will be influenced not only by others (friends, enemies) surrounding us but also by ourselves. My latter self will inevitably be influenced by my former self, and the weight of my own influence is usually positive.[5] Such kinds of setup of one's own influence can explain why some

---

[5] Only in some extreme cases, say a person encounters serious setbacks and loses his or her self-confidence, then his or her own influence could be negative.

people are hard to be influenced by others while some other people are easy to be influenced because the former individuals' own influences may have higher weights.

5 Social Influence Function: A Nonordering Approach

In a nonranked choice context, every agent can choose only his or her first preferred alternative but not present a full ordering among all alternatives in his or her ballot. As agents can only observe other agents' choices about their most preferred alternatives, the influences among agents can only work in a nonordering way, from the influencing agents' 1-of-m choices to the influenced agent's choice.[6] As illustrated by the example in Fig. 1, the influences flow from all influencing agents' 1-of-3 choices to agent M's choice. As each agent's choice is made according to his or her first preferred alternative, the influencing choice will be c for agent F_1, a for agent F_2, b for agent E, and a for agent M.

5.1 Plurality Influence Rule

The plurality rule (the alternative that obtains the highest number of votes wins) can be extended to a signed and weighted version referred to as a multi-influence aggregation method that is used to measure the weighted sum of the influence scores of each alternative obtained from all the influencing agents' choices.

Definition 5. *(Plurality Influence Score) is an influence score depending on how many times an alternative is most preferred by all influencing agents:*

$$I_o^P(i) = \sum_{O_1^j = o} w_{(j,i)} \quad i \in \mathbb{N}, o \in \mathbb{M}$$

where $I_o^P(i)$ represents the plurality score of influence on agent i obtained by alternative o and O_1^j indicates the first preferred alternative by agent $j \in \mathbb{N}$.

To compare the *plurality influence scores* among three alternatives in the example in Fig. 1, the sum of the weighted plurality scores of influence on agent M obtained by alternative a is $I_a^P(M) = w_{(F_2,M)} + w_{(M,M)} = 2 + 2 = 4$ (2 from agent F_2 and 2 from agent M), that obtained by alternative b is $I_b^P(M) = w_{(E,M)} = -1$ (from agent E), and that obtained by alternative c is $I_c^P(M) = w_{(F_1,M)} = 4$ (from agent F_1). Thus, both alternative a and c have the highest weighted score of influence (or, more briefly, weight of influence) on agent M according to the *plurality influence rule*, it is not clear for the result of the influenced choice for agent M.

---

[6] We cannot address the full preference orderings influencing and being influenced, as such information about orderings is inaccessible.

5.2 Majority Influence Rule

The majority rule (the alternative that obtains more than half of votes wins) can also be extended to a signed and weighted version referred to as a multi-influence aggregation method.[7]

Definition 6. *(Majority Influence Score) can be understood as the plurality influence score counted with more than one rounds, eliminating portion of alternatives each round, until an alternative obtains more than half of the total influence scores:* $\frac{\sum_{j\in N} w_{(j,i)}}{2}$.

If we use a *second-round majority influence rule*, then only alternative a and c can be compared in the second round of influence. The sum of the weighted majority scores of influence on agent M obtained by alternative a is $I_a^P(M) = w_{(F_2,M)} + w_{(M,M)} = 2 + 2 = 4$, and that obtained by alternative c is $I_c^P(M) = w_{(F_1,M)} + w_{(E,M)} = 4 - 1 = 3$ (4 from agent F_1 and -1 from agent E as alternative b has been eliminated). Thus, alternative a has the highest weighted score of influence on agent M according to the *majority influence rule*, which will be the result of the influenced choice for agent M.

6 Social Influence Function: An Ordering-Based Approach

In a ranked choice context, every agent can present a full ordering among all alternatives in their ballots. As agents can view other agents' full preference orderings of the alternatives, the influences among agents can work in an ordering way, from the influencing agents' preference orderings to the influenced agent's preference ordering. As illustrated by the example in Fig. 1, the influences flow directly from all influencing agents' preference orderings, including agent F_1's ordering $c \succ a \succ b$, agent F_2's ordering $a \succ c \succ b$, agent E's ordering $b \succ c \succ a$, and agent M's original preference ordering before the influence, $a \succ b \succ c$, to agent M's influenced preference ordering. To address the simultaneous influences of more than one agent in an ordering approach, new influence functions (rules) need to be built and developed.

6.1 Borda Influence Rule

The Borda count (allocating scores according to alternatives' ranks) can be extended to a signed and weighted version referred to as a multi-influence aggregation method that is used to measure the weighted sum of the influence scores of each alternative obtained from all the influencing agents' preference orderings.

---

[7] Actually, [11] also discussed a *majority influence rule* but just in a binary decision context.

Definition 7. *(Borda Influence Score) is an influence score depending on the ranks of an alternative in all influencing agents' preference orderings:*

$$I_o^B(i) = \sum_{O_r^j = o} (m - r + 1) \times w_{(j,i)} \quad i \in N, o \in M$$

where $I_o^B(i)$ represents the Borda score of influence on agent i obtained by alternative o, $r \in \{1, 2, ..., m\}$ represents the rank in an ordering, and O_r^j indicates the rth preferred alternative by agent $j \in N$.

To compare the *Borda influence scores* obtained by three alternatives in the example in Fig. 1, the following points are considered: (1) the Borda score of influence on agent M obtained by alternative a is $I_a^B(M) = w_{(F_1,M)} \times 1 + w_{(F_2,M)} \times 3 + w_{(E,M)} \times 1 + w_{(M,M)} \times 3 = 4 \times 1 + 2 \times 3 + (-1) \times 1 + 2 \times 3 = 15$ (in which 4×1 means the Borda score of influence of alternative a given by agent F_1 is 1, as a is least preferred by agent F_1, and the weight of influence from agent F_1 to agent M is 4); (2) the Borda score of influence on agent M obtained by alternative b is $I_b^B(M) = w_{(F_1,M)} \times 2 + w_{(F_2,M)} \times 1 + w_{(E,M)} \times 3 + w_{(M,M)} \times 2 = 4 \times 2 + 2 \times 1 + (-1) \times 3 + 2 \times 2 = 11$; (3) the Borda score of influence on agent M obtained by alternative c is $I_c^B(M) = w_{(F_1,M)} \times 3 + w_{(F_2,M)} \times 2 + w_{(E,M)} \times 2 + w_{(M,M)} \times 1 = 4 \times 3 + 2 \times 2 + (-1) \times 2 + 2 \times 1 = 16$. Thus, c has the highest weighted Borda score of influence on agent M, a has the second highest score, and b has the lowest score; then, according to the *Borda influence rule*, the result of the influenced preference for agent M will be $c \succ a \succ b$.

6.2 Condorcet Influence Rule

The Condorcet method (holding pairwise comparison between each of two alternatives) can also be extended to a signed and weighted version referred to as a multi-influence aggregation method that is used to measure the weighted sum of the influence score of each alternative obtained in every pairwise comparison with other alternatives of all the influencing agents' preference orderings.

Definition 8. *(Condorcet Influence Score) is a relative influence score of an alternative depending on which other alternative is to be compared in all influencing agents' preference orderings:*

$$I_{o \succ o'}^C(i) = \sum_{o \succ_j o'} w_{(j,i)} \quad i \in N, o, o' \in M$$

where $I_{o \succ o'}^C(i)$ represents the Condorcet score of influence on agent i obtained by alternative o compared with alternative o' and $o \succ_j o'$ indicates any agent $j \in N$ preferring alternative o to alternative o'.

To compare the pairwise *Condorcet influence scores* among three alternatives in the example in Fig. 1, there will be $C_3^2 = 3$ pairs of comparison: (1) a and b, the Condorcet score of influence on agent M obtained by alternative a compared with alternative b is $I^{\mathbf{C}}_{a \succ b}(M) = w_{(F_2,M)} + w_{(M,M)} = 2 + 2 = 4$ (in which the former 2 and the latter 2, respectively, mean that agent F_2 and agent M deem a better than b and both their weights are 2), and the Condorcet score of influence on agent M obtained by alternative b compared with alternative a is $I^{\mathbf{C}}_{b \succ a}(M) = w_{(F_1,M)} + w_{(E,M)} = 4 - 1 = 3$, $I^{\mathbf{C}}_{a \succ b}(M) > I^{\mathbf{C}}_{b \succ a}(M)$; thus, a has a stronger influence than b on agent M; (2) a and c, the Condorcet score of influence on agent M obtained by alternative a compared with alternative c is $I^{\mathbf{C}}_{a \succ c}(M) = w_{(F_2,M)} + w_{(M,M)} = 2 + 2 = 4$, and the Condorcet score of influence on agent M obtained by alternative c compared with alternative a is $I^{\mathbf{C}}_{c \succ a}(M) = w_{(F_1,M)} + w_{(E,M)} = 4 - 1 = 3$, $I^{\mathbf{C}}_{a \succ c}(M) > I^{\mathbf{C}}_{c \succ a}(M)$; thus, a also has a stronger influence than c on agent M; (3) b and c, the Condorcet score of influence on agent M obtained by alternative b compared with alternative c is $I^{\mathbf{C}}_{b \succ c}(M) = w_{(E,M)} + w_{(M,M)} = -1 + 2 = 1$, and the Condorcet score of influence on agent M obtained by alternative c compared with alternative b is $I^{\mathbf{C}}_{c \succ b}(M) = w_{(F_1,M)} + w_{(F_2,M)} = 4 + 2 = 6$, $I^{\mathbf{C}}_{b \succ c}(M) < I^{\mathbf{C}}_{c \succ b}(M)$; thus, c has a stronger influence than b on agent M. In conclusion, according to the *Condorcet influence rule*, alternative a can be named the *Condorcet influence winner* for agent M, as it can beat any other alternative in a pairwise comparison of the weighted score of influence, and a full preference ordering for agent M after being influenced will be $a \succ c \succ b$.

7 Matrix Influence Function: Application

Finally, we use the *matrix influence function* that is a signed and weighted version of the KSB metric [2,21,28,29] in the context of multi-influence; then, the feasible ordering that has the minimum weighted sum of distances from all influencing agents' preference orderings compared with all other feasible orderings should be chosen as the result of the influenced preference. Different from *social influence functions* we built in the above (such as the *Borda influence function* and the *Condorct influence function*) that focus on individual alternatives (such as finding which alternative has the highest score, or which alternative is pairwise preferred compared with all other alternatives), the *matrix influence function* directly handles full preference orderings.

To compare the distance between any two preference orderings, first, the ordering matrix OM for each feasible preference ordering (regardless of whether it is possessed by an agent already or just exists theoretically) should be given, as illustrated by the example in Fig. 1 and as shown in Table 1:

Further, based on the rule discussed above, it finds one possible preference p ($p \in \{a \succ b \succ c, a \succ c \succ b, b \succ a \succ c, b \succ c \succ a, c \succ a \succ b, c \succ b \succ a\}$) that has the minimum weighted sum of distances from all the influencing agents' preference orderings $(P_{(F_1)}, P_{(F_2)}, P_{(E)}, P_{(M)})$:

$$P'_{(M)} = \arg \min_{p} [4Dis(P_{(F_1)},p) + 2Dis(P_{(F_2)},p) - Dis(P_{(E)},p) + 2Dis(P_{(M)},p)]$$

There are both positive influence from a friend and negative influence from an enemy. Thus, the distance we compute here is not only weighted but also

Table 1. The ordering matrices of all feasible preference orderings

$$OM_{a \succ b \succ c} = OM_{P_{(M)}} = \begin{array}{c} \\ a \\ b \\ c \end{array} \begin{array}{ccc} a & b & c \\ \left(\begin{array}{ccc} 0 & -1 & -1 \\ 1 & 0 & -1 \\ 1 & 1 & 0 \end{array} \right) \end{array} \quad OM_{a \succ c \succ b} = OM_{P_{(F_2)}} = \begin{array}{c} \\ a \\ b \\ c \end{array} \begin{array}{ccc} a & b & c \\ \left(\begin{array}{ccc} 0 & -1 & -1 \\ 1 & 0 & 1 \\ 1 & -1 & 0 \end{array} \right) \end{array}$$

$$OM_{b \succ a \succ c} = \begin{array}{c} \\ a \\ b \\ c \end{array} \begin{array}{ccc} a & b & c \\ \left(\begin{array}{ccc} 0 & 1 & -1 \\ -1 & 0 & -1 \\ 1 & 1 & 0 \end{array} \right) \end{array} \quad OM_{b \succ c \succ a} = OM_{P_{(E)}} = \begin{array}{c} \\ a \\ b \\ c \end{array} \begin{array}{ccc} a & b & c \\ \left(\begin{array}{ccc} 0 & 1 & 1 \\ -1 & 0 & -1 \\ -1 & 1 & 0 \end{array} \right) \end{array}$$

$$OM_{c \succ a \succ b} = \begin{array}{c} \\ a \\ b \\ c \end{array} \begin{array}{ccc} a & b & c \\ \left(\begin{array}{ccc} 0 & -1 & 1 \\ 1 & 0 & 1 \\ -1 & -1 & 0 \end{array} \right) \end{array} \quad OM_{c \succ b \succ a} = OM_{P_{(F_1)}} = \begin{array}{c} \\ a \\ b \\ c \end{array} \begin{array}{ccc} a & b & c \\ \left(\begin{array}{ccc} 0 & 1 & 1 \\ -1 & 0 & 1 \\ -1 & -1 & 0 \end{array} \right) \end{array}$$

signed. It will play a role in finding the "closest" preference that is possible while computing the distance with positively influencing preferences and finding the "farthest" preference that is possible while computing the distance with negatively influencing preferences.

The computation outcome of all 6 possible orderings' weighted sums of distances from all influencing agents' preference orderings is shown in Table 2, and the result of the influenced preference ordering for agent M should be $a \succ c \succ b$, which possesses the minimum weighted sum of distances: $8 \times 4 + 0 \times 2 + 12 \times -1 + 4 \times 2 = 28$.

Table 2. Finding the result of the influenced preference by the matrix influence rule

	$c \succ_{F_1} b \succ_{F_1} a$ (4)	$a \succ_{F_2} c \succ_{F_2} b$ (2)	$b \succ_E c \succ_E a$ (−1)	$a \succ_M b \succ_M c$ (2)	Dis
$a \succ b \succ c$	12×4	4×2	8×-1	0×2	48
$a \succ c \succ b$	8×4	0×2	12×-1	4×2	28
$b \succ a \succ c$	8×4	8×2	4×-1	4×2	52
$b \succ c \succ a$	4×4	12×2	0×-1	8×2	56
$c \succ a \succ b$	4×4	4×2	8×-1	8×2	32
$c \succ b \succ a$	0×4	8×2	4×-1	12×2	36

8 Discussion, Conclusion and Future Work

We consider the scenario of group decisions where agents' decision-making behaviors and preferences can influence and be influenced by each other. To address the multiple influences among agents and obtain the preference of an agent after being influenced by others, previous work discussed the situation of influence of one agent (at a time) on another agent or the simultaneous influence of more than one agent in a utility, belief, opinion, or choice context, mainly using cardinal values-based approaches but not ordinal preference-based approaches. In addition, the variation in the signed and weighted individual influence has not been fully discussed. In this paper, we discuss how to address multiple sources of influence with varied strengths and opposite polarities in an ordering-based approach by extending classical *social choice functions* to signed and weighted *social influence functions*, such as the *Condorcet influence function* and the *Borda influence function*. In particular, we extend the KSB distance metric to a *matrix influence function* and define the rule regarding how to transform each preference ordering into a matrix (named an *ordering matrix*) and set a measure to compute the distance between any two orderings in the sense of influence. However, our work is not yet sufficient, as other prospects remain:

- We have discussed how to acquire the resulting choice or preference when we choose one or another *social influence function*. It is common that different *social influence functions* lead to different results. Thus, how to find a suitable benchmark to measure how good a *social influence function* is or what properties a good *social influence function* should possess needs to be discussed in the future, similar to the comparison and evaluation of *social choice functions* (group decision methods, voting methods, and so on).
- We have not provided enough material to support these *social influence functions* to accurately characterize influences among agents in practice. In the future, the contributions of other disciplines such as cognitive science and experimental psychology would be valuable.
- We have preliminarily considered the variable of time and the iteration of influence in a general *matrix influence function* but have not yet experimented with these concepts. In the future, agent-based simulation could be used to demonstrate the dynamics of multiperiod influences in group decision-making, particularly with consideration of the network topology and constraints.

Acknowledgment. This study is supported by a Natural Science Foundation of China Grant (71804006) and a National Natural Science Foundation of China and European Research Council Cooperation and Exchange Grant (7161101045).

References

1. Antal, T., Krapivsky, P.L., Redner, S.: Social balance on networks: the dynamics of friendship and enmity. Physica D Nonlinear Phenomena **224**(1), 130–136 (2006)
2. Bogart, K.P.: Preference structures I: distances between transitive preference relations? J. Math. Sociol. **3**(1), 49–67 (1973)
3. Degroot, M.H.: Reaching a consensus. J. Am. Stat. Assoc. **69**(345), 118–121 (1974)
4. Demarzo, P.M., Vayanos, D., Zwiebel, J.: Persuasion bias, social influence, and unidimensional opinions. Q. J. Econ. **118**(3), 909–968 (2003)

5. Golub, B., Jackson, M.O.: Naive learning in social networks and the wisdom of crowds. Am. Econ. J. Microecon. **2**(1), 112–149 (2010)
6. Grabisch, M., Rusinowska, A.: A model of influence in a social network. Theory Decis. **69**(1), 69–96 (2010)
7. Grabisch, M., Rusinowska, A.: A model of influence with an ordered set of possible actions. Theory Decis. **69**(4), 635–656 (2010)
8. Grabisch, M., Rusinowska, A.: Iterating influence between players in a social network. In: 16th Coalition Theory Network Workshop (2011)
9. Grabisch, M., Rusinowska, A.: Measuring influence in command games. Soc. Choice Welfare **33**(2), 177–209 (2009)
10. Grabisch, M., Rusinowska, A.: Influence functions, followers and command games. Games Econ. Behav. **72**(1), 123–138 (2011)
11. Grabisch, M., Rusinowska, A.: A model of influence based on aggregation function. Math. Soc. Sci. **66**(3), 316–330 (2013)
12. Grandi, U., Lorini, E., Perrussel, L.: Propositional opinion diffusion. In: International Conference on Autonomous Agents and Multiagent Systems, pp. 989–997 (2015)
13. Granovetter, M.: The strength of weak ties: a network theory revisited. Sociol. Theory **1**(6), 201–233 (1983)
14. Granovetter, M.: The impact of social structure on economic outcomes. J. Econ. Perspect. **19**(1), 33–50 (2005)
15. Granovetter, M.S.: The strength of weak ties. Am. J. Sociol. **78**(6), 1360–1380 (1973)
16. Hao, F., Ryan, P.Y.: Real-World Electronic Voting: Design, Analysis and Deployment. CRC Press, Boca Raton (2016)
17. Harary, F., et al.: On the notion of balance of a signed graph. Michigan Math. J. **2**(2), 143–146 (1953)
18. Heider, F.: Attitudes and cognitive organization. J. Psychol. **21**(1), 107 (1946)
19. Hoede, C., Bakker, R.: A theory of decisional power. J. Math. Sociol. **8**, 309–322 (1982)
20. Jackson, M.O.: Social and Economic Networks. Princeton University Press, Princeton (2008)
21. Kemeny, J.G., Snell, J.L.: Mathematical Models in the Social Sciences. The MIT Press, Cambridge (1972)
22. Krackhardt, D.: The strength of strong ties: the importance of philos in organizations. In: Nohria, N., Eccles, R.G. (eds.) Networks and Organizations: Structure, Form, and Action, pp. 216–239. Harvard Business School Press (1992)
23. Luo, H.: Agent-based modeling of the UN security council decision-making: with signed and weighted mutual-influence. In: Social Simulation Conference 2018 (2018)
24. Luo, H., Meng, Q.: Multi-agent simulation of SC reform and national game. World Econ. Polit. **6**, 136–155 (2013). in Chinese
25. Maran, A., Maudet, N., Pini, M.S., Rossi, F., Venable, K.B.: A framework for aggregating influenced CP-nets and its resistance to bribery. In: Twenty-Seventh AAAI Conference on Artificial Intelligence (2013)
26. Maudet, N., Pini, M.S., Venable, K.B., Rossi, F.: Influence and aggregation of preferences over combinatorial domains. In: International Conference on Autonomous Agents and Multiagent Systems (2012)
27. Salehi-Abari, A., Boutilier, C.: Empathetic social choice on social networks. In: International Conference on Autonomous Agents and Multi-agent Systems, pp. 693–700 (2014)

28. Wicker, A.W., Doyle, J.: Interest-matching comparisons using CP-nets. In: Twenty-Second AAAI Conference on Artificial Intelligence (2007)
29. Wicker, A.W., Doyle, J.: Comparing preferences expressed by CP-networks. In: AAAI Workshop on Advances in Preference Handling, pp. 128–133 (2008)

Choosing a Committee Under Majority Voting

Fatma Aslan[1,2]($\boxtimes$) (iD), Hayrullah Dindar[2] (iD), and Jean Lainé[1,3] (iD)

[1] CNAM-LIRSA, Paris, France
fatma.aslan@bilgi.edu.tr
[2] Istanbul Bilgi University, Istanbul, Turkey
[3] Murat Sertel Center for Advanced Economic Studies, Istanbul, Turkey

Abstract. We consider the elections of a seat-posted committee, and investigate the propensity of seat-wise majority voting to choose a committee that fulfills the majority will with respect to preferences over committees. Voters have seat-wise preferences and preferences over committees are derived from seat-wise preferences by means of a neutral preference extension. Neutrality means that the names of candidates do not play any role. The majority committee paradox refers to a situation where a Condorcet winner exists for each seat, and a Condorcet winner committee also exists but does not coincide with the combination of seat-wise Condorcet winners. The majority committee weak paradox refers to a situation where the combination of seat-wise Condorcet winners is not a Condorcet winner among committees. We characterize the domains of preference extensions immune to each of the paradoxes.

Keywords: Committee election · Voting paradoxes · Majority voting · Separable preferences

JEL Class: D 71

Arrovian social choice theory provides a theoretical framework for evaluating social choice functions, which aggregates individual ordinal preferences over social alternatives, or candidates, into a collective outcome. In the case where the outcome is a single candidate, asking voters to report their ranking of candidates is not problematic. However, if a committee of several candidates is to be chosen, this informational requirement is hardly implementable in practice. Consider an election of a faculty council involving a dean, a vice-dean for research and a vice-dean for teaching. If there are four candidates per seat, fully expressing preferences means ranking the 64 possible outcomes. Clearly, as the number of seats or the number of candidates for each seat increase, referring to Arrovian social choice functions becomes less and less useful in practice. Designing a seat-wise procedure is a frequent solution that overcomes this difficulty. In a seat-wise

The first author is grateful to the CNRS PICS program 08001 (Université de Caen Normandie and Istanbul Bilgi University) for its financial support.

© Springer Nature Switzerland AG 2019
D. C. Morais et al. (Eds.): GDN 2019, LNBIP 351, pp. 33–42, 2019.
https://doi.org/10.1007/978-3-030-21711-2_3

procedure, voters report their preferences over candidates seat-wise, and candidates are selected seat-wise. It is well-known that a seat-wise procedure may not lead to the outcome that would prevail for a direct choice procedure where voters report their preferences over the outcomes. This happens when individual preferences exhibit complementarities among candidates, but this may even prevail with separable preferences which prohibits any sort of complementarity.

The potential inconsistency between seat-wise and direct procedures results from the fact that seat-wise preferences describe only partially preferences over outcomes. A rather rich literature dealing with this inconsistency and other potential drawbacks of seat-wise procedures deals with multiple referenda, which is equivalent to a committee choice problem with two candidates per seat. In this setting, each voter is characterized by an ideal committee, and simple majority voting provides a natural seat-wise choice procedure that we denote by Maj. Compound majority voting paradoxes studied in the literature express the fact that Maj may lead to outcomes exhibiting some undesirable properties. The Anscombe's paradox (Anscombe 1976; Wagner 1984; Laffond and Lainé 2013) shows that a majority of voters may disagree with the outcome of Maj on a majority of seats. The multiple elections paradox (Brams et al. 1998; Scarsini 1998) prevails when the winner for Maj receives zero votes in the direct elections (or, equivalently, may be ranked first by no voter). The Ostrogorski paradox (Ostrogorski 1902; Rae and Daudt 1976; Bezembinder and Van Acker 1985; Deb and Kelsey 1987; Kelly 1989; Shelley 1994; Laffond and Lainé 2006) prevails when another outcome beats the one of Maj according to majority voting under the assumption that committees are compared by means of the Hamming distance criterion.[1,2]

The Hamming distance criterion provides a specific way to relate seat-wise preferences and preferences over committees. Other ways can be considered, each referring to a particular preference extension. Formally, when there are only two candidates per seat, a preference extension rule maps each ideal committee to a (weak) ordering of committees. A usual property retained for a preference extension is separability: if a and b are the two candidates for some seat s, and if a voter ranks a above b, she will rank two committees identical in all seats but s according to her preference over a and b.[3] Kadane (1972) shows that even under the assumption of a separable extension, Maj may select a Pareto

---

[1] Hamming distance criterion in this specific setting simply means that voters prefer the committee(s) agreeing with her ideal on a higher number of seats.

[2] Laffond and Lainé (2009) show that Maj always selects a Pareto optimal element in the Top-Cycle of the majority tournament among outcomes (Schwartz 1972) while Maj may select an outcome which does not belong to the Uncovered set (Miller (1977); Moulin (1986)). An overview of compound majority paradoxes in multiple referenda is provided in Laffond and Lainé (2010).

[3] Lacy and Niou (2000) show that under a non-separable preference extension rule, Maj may select a Condorcet-loser outcome (i.e., an outcome majority defeated by all other outcomes). However, if separability holds, Maj always chooses the Condorcet winner outcome (i.e., the outcome majority defeating all other outcomes) whenever it exists (Kadane 1972).

dominated committee. Moreover, Özkal-Sanver and Sanver (2006) show that if there are at least three seats, no anonymous seat-wise procedure guarantees that a Pareto optimal committee will be chosen for all separable preference extensions. However, for the Hamming preference extension rule, Maj always produce a Pareto-optimal committee (Brams et al. 2007). Çuhadaroğlu and Lainé (2012) prove that under a mild richness assumption, the Hamming preference extension defines the largest domain of separable preference extensions for which Maj always picks a Pareto optimal outcome.[4]

All the above mentioned studies deal with the case of two candidates per seat. Less attention has been paid to situations where there are more than two candidates per seat. Benoît and Kornhauser (2010) generalize the result of Özkal-Sanver and Sanver (2006): if any separable preference extension is admissible, and if there are at least three seats or when there are precisely two seats with more than two candidates per seat, a seat-wise procedure selects a Pareto optimal outcome if and only if it is dictatorial. While this strong result disqualifies seat-wise procedures in the full domain of separable preferences extensions, it suggests investigating whether they can perform better under some domain restrictions. This is the route followed in this paper, which addresses the following question: can we characterize the class of preference extensions under which Maj selects a Condorcet winner committee, that is a committee preferred to all other committees by a majority of voters?

One difficulty is that Maj is not well-defined with at least three candidates per seat. Indeed, it is well-known that a Condorcet winner for each seat (i.e., a candidate preferred to all other candidates by a majority of voters) may fail to exist. However, well-known restrictions upon voters' preferences ensure the existence of a seat-wise Condorcet winner are single-peakedness and Sen's value restriction (Black 1948; Sen 1966). We assume that preferences over candidates for each seat are such that a Condorcet winner exists, and we address the following problem: characterizing the preference extension domain for which the committee formed by all seat-wise Condorcet winners form a Condorcet winner among committees. If Maj selects a Condorcet winner committee, there is no inconsistency between seat-wise majority voting and direct majority voting (where voters rank committees). Hence, characterizing the preference extension domain that precludes this inconsistency solves the problem created by the Arrovian informational requirement: in order to fulfill the majority will for committees, it is sufficient to fulfill the majority will for each seat.

The inconsistency between seat-wise majority voting and direct majority voting arises in two cases, each related to a new voting paradox. The *majority committee paradox* prevails when a Condorcet winner committee exists and is not selected by Maj. The *majority committee weak paradox* prevails when either the majority committee paradox holds or a Condorcet winner committee fails to exist (while Maj is well-defined).

---

[4] Within a similar setting, Laffond and Lainé (2012) show that Maj may fail at implementing a compromise, even under strong restrictions upon the seat-wise majority margin.

Under a neutrality assumption for preference extensions (meaning that candidates' names play no role), we characterize the preference extension domain immune to the majority committee paradox and the one immune to the majority committee weak paradox in the case of two-seat committees. More precisely, we prove that separability is a necessary and sufficient condition for a neutral preference extension to avoid the majority committee paradox. Moreover, the domain of neutral preference extensions avoiding the majority committee weak paradox is much smaller, it reducing to a unique lexicographic preference extension. According to lexicographic preference extensions all voters agree on a seat as priority seat and compares committees according to their ranking of candidates for that priority seat whenever they differ and if both committees have the same candidate on the priority seat, compares them according to ranking of candidates for other seat.

Our results complement the ones obtained by Hollard and Le Breton (1996) and Vidu (1999, 2002). In the case of two candidates per seat, Hollard and Le Breton (1996) show that any separable tournament over committees can be achieved through seat-wise majority voting. This result is generalized in Vidu (1999) to the case of more than two candidates per seat. Moreover, Vidu (2002) shows that a similar result prevails even when seat-wise preferences are single-peaked (implying the existence of seat-wise Condorcet winners).

The paper is organized as follows. Section 2 provides basic notations and definitions. Majority committee paradoxes are formalized in Sect. 3. Results are stated in Sect. 4. We conclude with several comments about further research.

1   The Model

1.1   Preliminaries

We consider two finite sets $\mathbb{C}_1$ and $\mathbb{C}_2$, with respective cardinalities C_1 and C_2. Sets $\mathbb{C}_1$ and $\mathbb{C}_2$ are interpreted as sets of candidates competing for seat-1 and for seat-2. We use letters a, b, c to denote arbitrary candidates for seat-1 and x, y, z to denote arbitrary candidates for seat-2. A *committee* $\mathcal{C}$ is an element of $\mathbb{C} = \mathbb{C}_1 \times \mathbb{C}_2$.

We also consider a finite set of voters $\mathcal{N} = \{1, ..., n, ..., N\}$ where $N \in \mathbb{N}$ is odd. For each $t \in \{1, 2\}$, every voter n has preferences over candidates in $\mathbb{C}_t$, called t-preferences, represented by a complete linear order P_n^t. The upper-contour set of a candidate $a \in \mathbb{C}_t$ for P_n^t is defined by $U(a, P_n^t) = \{b \in \mathbb{C}_t : bP_n^t a\}$. Moreover, the rank of a in P_n^t is defined by $r^t(a, P_n^t) = 1 + |U(a, P_n^t)|$. Given a finite set X, let $\mathcal{L}(X)$ denote the set of linear orders over X. Finally, for $t = 1, 2$, a t-profile $\pi^t = (P_n^t)_{n \in \mathcal{N}}$ is an element of $(\mathcal{L}(\mathbb{C}_t))^N$.

We call preference over committees, or in short preference, of voter n an element $P_n = (P_n^1, P_n^2)$ of $\mathcal{L}(\mathbb{C}_1) \times \mathcal{L}(\mathbb{C}_2)$. The P_n-rank vector of committee $\mathcal{C} = (a, x) \in \mathbb{C}$ is defined by $r(\mathcal{C}, P) = (r^1(a, P_n^1), r^2(x, P_n^2))$. A *profile* is an element $\pi = (P_n)_{n \in \mathcal{N}}$ of $(\mathcal{L}(\mathbb{C}_1) \times \mathcal{L}(\mathbb{C}_2))^N$.

1.2 Preference Extension

Seat-wise preferences over candidates and preferences are logically related. We assume that preferences over candidates for each seat are extended to preferences by means of a preference extension. Formally, *a preference extension is a mapping* δ *from* $\mathcal{L}(\mathbb{C}_1) \times \mathcal{L}(\mathbb{C}_2)$ *to* $\mathcal{L}(\mathbb{C})$. *A preference extension profile* is a vector $\delta^{\mathcal{N}} = (\delta_1, ..., \delta_N)$ of preference extensions. Given a profile $\pi = (P_n)_{n \in \mathcal{N}} \in (\mathcal{L}(\mathbb{C}_1) \times \mathcal{L}(\mathbb{C}_2))^N$, a preference extension profile $\delta^{\mathcal{N}}$ generates the extended profile $\delta^{\mathcal{N}}(\pi) = ((\delta_n(P_n))_{n \in \mathcal{N}} \in (\mathcal{L}(\mathbb{C}))^N$.

We retain two properties for preference extensions, neutrality and separability. Neutrality prevails if the names of candidates do not matter when comparing committees. In other words, only ranks given to candidates are taken into account.

Definition 1. *A preference extension* δ *is* **neutral** *if for all* $P = (P^1, P^2)$, $P' = (P'^1, P'^2) \in \mathcal{L}(\mathbb{C}_1) \times \mathcal{L}(\mathbb{C}_2)$, *and for all* $C, C' \in \mathbb{C}$, *if* $[r(C, P) = r(C, P')$ *and* $r(C', P) = r(C', P')]$ *then* $[C\delta(P)C' \Leftrightarrow C\delta(P')C']$.

It follows from Definition 1 that given a preference $P \in \mathcal{L}(\mathbb{C}_1) \times \mathcal{L}(\mathbb{C}_2)$, any neutral preference extension δ can be equivalently defined as the linear order $\gg_{\delta}$ over $\{1, ..., C_1\} \times \{1, ..., C_2\}$ by: for all $i, j \in \{1, ..., C_1\} \times \{1, ..., C_2\}$, $i \gg_{\delta} j$ if and only if there exists $C, C' \in \mathbb{C}$ such that $C\delta(P)C'$ where $r(C, P) = i$ and $r(C', P) = j$. Hence, *a neutral preference extension profile* can be equivalently defined by vector $\gg_{\delta^{\mathcal{N}}} = (\gg_{\delta_1}, ..., \gg_{\delta_N})$.

We denote the set of neutral preference extensions by Δ.

The following example illustrates how a neutral extension operates. Let $\mathbb{C}_1 = \{a, b\}$, $\mathbb{C}_2 = \{x, y\}$, $\mathcal{N} = \{1, 2, 3\}$, and consider the seat-wise profiles $\pi = (\pi^1, \pi^2) = ((P_n^1, P_n^2)_{n=1,2,3})$ defined below:

$$\pi^1 = \begin{pmatrix} 1 & 2 & 3 \\ \hline a & b & a \\ b & a & b \end{pmatrix}, \quad \pi^2 = \begin{pmatrix} 1 & 2 & 3 \\ \hline x & y & x \\ y & x & y \end{pmatrix}$$

Let $\gg_{\delta^{\mathcal{N}}}$ be the following neutral preference extension profile:

$$\gg_{\delta^{\mathcal{N}}} = \begin{pmatrix} \gg_{\delta_1} & \gg_{\delta_2} & \gg_{\delta_3} \\ \hline (1,1) & (1,1) & (1,1) \\ (2,1) & (2,1) & (1,2) \\ (1,2) & (1,2) & (2,1) \\ (2,2) & (2,2) & (2,2) \end{pmatrix}$$

$\gg_{\delta^{\mathcal{N}}}$ combined with $\pi = (\pi^1, \pi^2)$ lead to the following extended profile:

$$\delta^{\mathcal{N}}(\pi) = \begin{pmatrix} 1 & 2 & 3 \\ \hline (a,x) & (b,y) & (a,x) \\ (b,x) & (a,y) & (a,y) \\ (a,y) & (b,x) & (b,x) \\ (b,y) & (a,x) & (b,y) \end{pmatrix}$$

Hereafter we refer to neutral preference extensions simply as preference extensions. The second property for preference extension rules we consider, called separability, holds if it is always preferable to assign a seat to a better candidate whoever the other committee member is. Hence, separability precludes any complementarity between candidates for different seats.

Definition 2. *A preference extension δ is separable if for all $a, b \in \mathbb{C}_1$, for all $x, y \in \mathbb{C}_2$, and for all $P = (P^1, P^2) \in \mathcal{L}(\mathbb{C}_1) \times \mathcal{L}(\mathbb{C}_2)$,*

- *$(a, x)\ \delta(P)\ (b, x)$ if and only if $a\ P^1\ b$*
- *$(a, x)\ \delta(P)\ (a, y)$ if and only if $x\ P^2\ y$.*

Under neutrality, separability is equivalently defined as follows:

- for all $(i_1, i_2) \neq (j_1, i_2) \in \{1, ..., C_1\} \times \{1, ..., C_2\}$, $(i_1, i_2) \gg_\delta (j_1, i_2)$ if $i_1 < j_1$
- for all $(i_1, i_2) \neq (i_1, j_2) \in \{1, ..., C_1\} \times \{1, ..., C_2\}$, $(i_1, i_2) \gg_\delta (i_1, j_2)$ if $i_2 < j_2$.

The set of neutral and separable extensions is denoted by Δ^{sep}. We introduce below two specific elements of this set:

Definition 3.

- *The 1-lexicographic preference extension δ^{1Lex} is defined by:*
  $(1, 1) \gg (1, 2) \gg ... \gg (1, C_2) \gg ... \gg (C_1, 1) \gg (C_1, 2)... \gg (C_1, C_2)$.
- *The 2-lexicographic preference extension, δ^{2Lex} is defined by:*
  $(1, 1) \gg (2, 1) \gg ... \gg (C_1, 1) \gg ... \gg (1, C_2) \gg (2, C_2)... \gg (C_1, C_2)$.

The following definition of a choice problem summarizes all the relevant features of a committee selection procedure:

Definition 4. *A choice problem is a 5-tuple $\mathcal{P} = (\mathbb{C}_1, \mathbb{C}_2, \mathcal{N}, \pi, \delta^\mathcal{N})$ where $\mathbb{C}_1$ and $\mathbb{C}_2$ are the set of candidates for seat-1 and seat-2 respectively, $\mathcal{N}$ is the set of voters with profile $\pi \in (\mathcal{L}(\mathbb{C}_1) \times \mathcal{L}(\mathbb{C}_2))^N$, and $\delta^\mathcal{N} \in \Delta^N$ is a preference extension profile.*

2   Majority Voting Paradoxes

We now formalize seat-wise and direct selection procedures based on simple majority voting. This requires formalizing two types of majority tournaments. Given $t \in \{1, 2\}$ together with a t-profile $\pi^t = (P_n^t)_{n \in \mathcal{N}} \in (\mathcal{L}(\mathbb{C}_t))^N$, the π^t−majority tournament is the complete and asymmetric binary relation $T(\pi^t)$ defined over $\mathbb{C}_t \times \mathbb{C}_t$ by: $\forall a, b \in \mathbb{C}_t$, $aT(\pi^t)b$ if and only if $|\{n \in \mathcal{N} : xP_n^t y\}| > \frac{N}{2}$. If $aT(\pi^t)b$, we say that candidate a defeats candidate b in π^t. Similarly, given a profile $\pi = (P_n)_{n \in \mathcal{N}} \in (\mathcal{L}(\mathbb{C}_1) \times \mathcal{L}(\mathbb{C}_2))^N$ together with a preference extension profile $\delta^\mathcal{N}$, the $\delta^\mathcal{N}(\pi)$−majority tournament is the complete and asymmetric binary relation $T(\delta^\mathcal{N}(\pi))$ defined over $\mathbb{C} \times \mathbb{C}$ by: $\forall\ \mathcal{C}, \mathcal{C}' \in \mathbb{C}$, $\mathcal{C}T(\delta^\mathcal{N}(\pi))\mathcal{C}'$ if $|\{n \in \mathbb{N} : \mathcal{C}\ \delta_n(P_n)\ \mathcal{C}'\}| > \frac{N}{2}$. If $\mathcal{C}T(\delta^\mathcal{N}(\pi))\mathcal{C}'$, we say that committee $\mathcal{C}$ defeats committee $\mathcal{C}'$ in $\delta^{\mathcal{N}}(\pi)$. Moreover, $T(\pi^t)$ (resp. $T(\delta^{\mathcal{N}}(\pi))$) admits a (necessarily unique) Condorcet winner if there exists a candidate $c(T(\pi^t)) \in \mathbb{C}_t$ (resp. a committee $c(T(\delta^{\mathcal{N}}(\pi)))$) that defeats all other candidates in π^t (resp. in $\delta^{\mathcal{N}}(\pi)$). We adopt the convention $c(T(\pi^t)) = \varnothing$ (resp. $c(T(\delta^{\mathcal{N}}(\pi))) = \varnothing$) when the underlying tournament has no Condorcet winner.

The seat-wise procedure consists of selecting a candidate for each seat from the seat-wise majority tournaments $T(\pi^1)$ and $T(\pi^2)$. We assume that preferences over candidates are restricted so as to ensure that both tournaments $T(\pi^1)$ and $T(\pi^2)$ admit a Condorcet winner. The direct procedure on the other hand consists of selecting a committee from the majority tournament over committees $T(\delta^{\mathcal{N}}(\pi))$. The seat-wise and direct procedures are inconsistent if either there is a Condorcet winner among committees that is not the combination of seat-wise Condorcet winners, or if there is no Condorcet winner among committees. This leads to the following two definitions of voting paradoxes:

Definition 5. *The majority committee paradox occurs at choice problem $\mathcal{P}$ if and only if $T(\pi_1)$, $T(\pi_2)$, and $T(\delta^{\mathcal{N}}(\pi))$ each admit a Condorcet winner while $c(T(\pi_1)) \times c(T(\pi_2)) \neq c(T(\delta^{\mathcal{N}}(\pi)))$.*

The majority committee paradox is illustrated in the following simple example:

Example 1. Let $\mathbb{C}_1 = \{a, b\}$, $\mathbb{C}_2 = \{x, y\}$, $\mathcal{N} = \{1, 2, 3\}$, and consider the seat-wise profiles $\pi = (\pi^1, \pi^2)$ defined below:

$$
\pi^1 = \begin{pmatrix} 1\ 2\ 3 \\ \hline a\ a\ a \\ b\ b\ b \end{pmatrix}, \pi^2 = \begin{pmatrix} 1\ 2\ 3 \\ \hline x\ x\ y \\ y\ y\ x \end{pmatrix}
$$

For $i = 1, 2, 3$ let $\gg_{\delta_i}$ be $(2, 1) \gg_{\delta_i} (1, 2) \gg_{\delta_i} (1, 1) \gg_{\delta_i} (2, 2)$, combined with π, leading to the following extended profile:

$$
\delta^{\mathcal{N}}(\pi) = \begin{pmatrix} 1 & 2 & 3 \\ \hline (b, x) & (b, x) & (b, y) \\ (a, y) & (a, y) & (a, x) \\ (a, x) & (a, x) & (a, y) \\ (b, y) & (b, y) & (b, x) \end{pmatrix}
$$

Clearly, $c(T(\pi^1)) \times c(T(\pi^2)) = (a, x)$ and $c(T(\delta^{\mathcal{N}}(\pi))) = (b, x)$, thus the majority committee paradox holds.[5]

Definition 6. *The majority committee weak paradox occurs at choice problem $\mathcal{P}$ if and only if $T(\pi_1)$ and $T(\pi_2)$ both admit a Condorcet winner while $c(T(\pi_1)) \times c(T(\pi_2)) \neq c(T(\delta^{\mathcal{N}}(\pi)))$.*

---

[5] Note that the preference extensions used by the voters are not separable which turns out to be necessary and sufficient to avoid the majority committee paradox as will be shown in Theorem 1.

The majority paradox implies the weak majority paradox, while the opposite is not true. Moreover, it is straightforward to show that the majority paradox never prevails when there are two candidates per seat.

The next example will illustrate the majority committee weak paradox.

Example 2. Let $\mathbb{C}_1 = \{a, b\}$, $\mathbb{C}_2 = \{x, y\}$, $\mathcal{N} = \{1, 2, 3\}$, and consider the seatwise profiles $\pi = (\pi^1, \pi^2)$ defined below:

$$
\pi^1 = \begin{pmatrix} 1\ 2\ 3 \\ \hline a\ b\ a \\ b\ a\ b \end{pmatrix}, \pi^2 = \begin{pmatrix} 1\ 2\ 3 \\ \hline x\ y\ y \\ y\ x\ x \end{pmatrix}
$$

Let $\gg_{\delta^{\mathcal{N}}}$ be such that

- $(1,1) \gg_{\delta_i} (2,1) \gg_{\delta_i} (1,2) \gg_{\delta_i} (2,2)$ for $i = 1, 3$.
- $(1,1) \gg_{\delta_3} (1,2) \gg_{\delta_3} (2,1) \gg_{\delta_3} (2,2)$.

Combination of $\gg_{\delta^{\mathcal{N}}}$ with π leads to the following extended profile:

$$
\delta^{\mathcal{N}}(\pi) = \begin{pmatrix} 1 & 2 & 3 \\ \hline (a,x) & (b,y) & (a,y) \\ (b,x) & (b,x) & (b,y) \\ (a,y) & (a,y) & (a,x) \\ (b,y) & (a,x) & (b,x) \end{pmatrix}
$$

(a, x) defeats (b, x), (b, x) defeats (a, y), (a, y) defeats (b, y), and finally, (b, y) defeats (a, x) in $\delta^{\mathcal{N}}(\pi)$, So, there is no Condorcet winner among committees while π_1 and π_2 both admit condorcet winners. Thus, majority committee weak paradox holds.[6]

3   Results

We define an extension domain as a non-empty subset $\mathcal{D}$ of Δ. First, we characterize the domain of preference extensions that are immune to the majority committee paradox in the following sense: $\mathcal{D} \subseteq \Delta$ is immune to majority committee paradox if and only if for all choice problems $\mathcal{P} = (\mathbb{C}_1, \mathbb{C}_2, \mathcal{N}, \pi, \delta^{\mathcal{N}})$ such that $c(T(\pi_1)) \neq \varnothing$, $c(T(\pi_2)) \neq \varnothing$ and $c(T(\delta^{\mathcal{N}}(\pi))) \neq \varnothing$; $\delta^{\mathcal{N}} \subseteq \mathcal{D}^N$ implies $c(T(\pi_1)) \times c(T(\pi_2)) = c(T(\delta^{\mathcal{N}}(\pi)))$. That is, a domain of preference extensions is immune to the majority committee paradox if and only if at any choice problem where all the voters use a preference extension from this domain and both seatwise and committee-wise Condorcet winners exist; the combination of seat-wise Condorcet winners is the Condorcet winning committee.

Theorem 1. *A preference extension domain $\mathcal{D}$ is immune to the majority paradox if and only if $\mathcal{D} \subseteq \Delta^{sep}$.*

---

[6] Note that for each voter the preference extension used is either δ^{1Lex} or δ^{2Lex}, but voters are not unanimously using one or the other which makes a significant difference as we will show in Theorem 2.

Our second result is a similar characterization for the weak majority weak paradox. Following an almost identical construction, we characterize the domain of preference extensions that are immune to the majority committee weak paradox in the following sense: $\mathcal{D} \subseteq \Delta$ is immune to majority committee paradox if and only if for all choice problems $\mathcal{P} = (\mathbb{C}_1, \mathbb{C}_2, \mathcal{N}, \pi, \delta^{\mathcal{N}})$ such that $c(T(\pi_1)) \neq \varnothing$ and $c(T(\pi_2)) \neq \varnothing$; $\delta^{\mathcal{N}} \subseteq \mathcal{D}^N$ implies $c(T(\pi_1)) \times c(T(\pi_2)) = c(T(\delta^{\mathcal{N}}(\pi)))$. That is, a domain of preference extensions is immune to the majority committee weak paradox if and only if at any choice problem where all the voters use a preference extension from this domain and both seat-wise Condorcet winners exist; Condorcet winning committee exists and is equal to the combination of seat-wise Condorcet winners.

Theorem 2. *A preference extension domain $\mathcal{D}$ is immune to the majority weak paradox if and only if either $\mathcal{D} = \{\delta^{1Lex}\}$ or $\mathcal{D} = \{\delta^{2Lex}\}$.*

4   Further Comments

Two routes are open for further research. The first is considering committee choice problems involving more than two seats. We strongly conjecture that results similar to Theorems 1 and 2 hold in this general setting. Finally, it would be interesting to characterize the domain of neutral preference extensions ensuring the existence of a Condorcet winning committee under the assumption that seat-wise Condorcet winners exist, disregarding whether the former is combination of latter ones.

References

Anscombe, G.E.M.: On frustration of the majority by fulfilment of the majority's will. Analysis **36**(4), 161–168 (1976)

Benoît, J.-P., Kornhauser, L.A.: Only a dictatorship is efficient. Games Econ. Behav. **70**(2), 261–270 (2010)

Bezembinder, T., Van Acker, P.: The Ostrogorski paradox and its relation to nontransitive choice. J. Math. Soc. **11**(2), 131–158 (1985)

Black, D.: On the rationale of group decision-making. J. Polit. Econ. **56**(1), 23–34 (1948)

Brams, S.J., Kilgour, D.M., Sanver, M.R.: A minimax procedure for electing committees. Publ. Choice **132**(3–4), 401–420 (2007)

Brams, S.J., Kilgour, D.M., Zwicker, W.S.: The paradox of multiple elections. Soc. Choice Welfare **15**(2), 211–236 (1998)

Çuhadaroğlu, T., Lainé, J.: Pareto efficiency in multiple referendum. Theory Decis. **72**(4), 525–536 (2012)

Deb, R., Kelsey, D.: On constructing a generalized Ostrogorski paradox: necessary and sufficient conditions. Math. Soc. Sci. **14**(2), 161–174 (1987)

Hollard, G., Breton, M.: Logrolling and a McGarvey theorem for separable tournaments. Soc. Choice Welfare **13**(4), 451–455 (1996)

Kadane, J.B.: On division of the question. Publ. Choice **13**(1), 47–54 (1972)

Kelly, J.S.: The Ostrogorski paradox. Soc. Choice Welfare **6**(1), 71–76 (1989)

Lacy, D., Niou, E.M.S.: A problem with referendums. J. Theor. Polit. **12**(1), 5–31 (2000)

Laffond, G., Lainé, J.: Single-switch preferences and the Ostrogorski paradox. Math. Soc. Sci. **52**(1), 49–66 (2006)

Laffond, G., Lainé, J.: Condorcet choice and the Ostrogorski paradox. Soc. Choice Welfare **32**(2), 317–333 (2009)

Laffond, G., Lainé, J.: Does choosing committees from approval balloting fulfill the electorate's will? In: Laslier, J.F., Sanver, M. (eds.) Handbook on Approval Voting. Studies in Choice and Welfare, pp. 125–150. Springer, Heidelberg (2010). https://doi.org/10.1007/978-3-642-02839-7_7

Laffond, G., Lainé, J.: Searching for a compromise in multiple referendum. Group Decis. Negot. **21**(4), 551–569 (2012)

Laffond, G., Lainé, J.: Unanimity and the Anscombe's paradox. Top **21**(3), 590–611 (2013)

Miller, N.: Graph theoretical approaches to the theory of voting. Am. J. Pol. Sci. **21**, 769–803 (1977)

Moulin, H.: Choosing from a tournament. Soc. Choice Welfare **2**, 271–291 (1986)

Ostrogorski, M.: Democracy and the Organization of Political Parties, vol. 2. Macmillan, London (1902)

Özkal-Sanver, I., Sanver, M.R.: Ensuring Pareto optimality by referendum voting. Soc. Choice Welfare **27**(1), 211–219 (2006)

Rae, D.W., Daudt, H.: The Ostrogorski paradox: a peculiarity of compound majority decision. Eur. J. Polit. Res. **4**(4), 391–398 (1976)

Scarsini, M.: A strong paradox of multiple elections. Soc. Choice Welfare **15**(2), 237–238 (1998)

Schwartz, T.: Rationality and the myth of the maximum. Noûs **6**, 97–117 (1972)

Sen, A.K.: A possibility theorem on majority decisions. Econometrica: J. Econom. Soc. **34**, 491–499 (1966)

Shelley, F.M.: Notes on Ostrogorski's paradox. Theory Decis. **17**(3), 267–273 (1994)

Vidu, L.: An extension of a theorem on the aggregation of separable preferences. Soc. Choice Welfare **16**(1), 159–167 (1999)

Vidu, L.: Majority cycles in a multi-dimensional setting. Econ. Theory **20**(2), 373–386 (2002)

Wagner, C.: Avoiding Anscombe's paradox. Theory Decis. **16**(3), 233–238 (1984)

Reciprocity and Rule Preferences of a Rotating Savings and Credit Association (ROSCA) in China: Evolutionary Simulation in Imitation Games

Zhao Sijia$^{(\boxtimes)}$ ⓘ and Masahide Horita ⓘ

Department of International Studies, Graduate School of Frontier Sciences,
The University of Tokyo, Kashiwa, Japan
rochester2045@gmail.com

Abstract. As one method of cooperation in human society, informal financial institutions, such as a ROSCA, demonstrate huge rule disparities temporally and geographically. In this paper, we attempt to understand whether and how people's preference of a ROSCA is related to the reciprocity level in a particular society. After conducting evolutionary imitation games among the population, the results show that each ROSCA rule evolves as if it finds its niche formed by the peoples' different levels of reciprocity. Our simulation also reproduced the social states where different ROSCAs co-exist with others at an equilibrium even when some rules clearly dominate others. These results provide a new insight into the theory of collective rule choice that triggers the evolution of cooperation.

Keywords: Cooperation · ROSCA · Rule preference · Reciprocity ·
Imitation game

1 Introduction

Based on documented history, we see that humans tend to cooperate by various ways. The study of the evolution of cooperation has attracted attention from a wide range of academic disciplines. More specifically, the process of human cooperation has been widely discussed. Trivers (1971) proposed that direct reciprocity promotes cooperation between a dyad of players interacting repeatedly, which was formalized as the repeated prisoner's dilemma game by Axelrod and Hamilton (1981). Nowak (2006) considered cooperation as a decisive organizing principle of human society, where evolution is found necessary to construct new levels of organization; then he discussed the novel five mechanisms for the evolution of cooperation, known as kin selection, direct reciprocity, indirect reciprocity, network reciprocity, and group selection.

Nowak and Sigmund (2005) stated that in human society, once cooperation is established, "a complex evolution takes place, which depends on the size of the population, the cost-to-benefit ratio, the average number of rounds, and the probability of errors." Among many factors affecting the evolution process, it is well acknowledged that reciprocity is of great significance, which leads to the establishment of

© Springer Nature Switzerland AG 2019
D. C. Morais et al. (Eds.): GDN 2019, LNBIP 351, pp. 43–56, 2019.
https://doi.org/10.1007/978-3-030-21711-2_4

cooperation (especially indirect cooperation). Although the subject is of much relevance to group decision-making, there has been little effort to investigate systematically the evolution of cooperation by practical cases to reveal more evidence on this topic.

A rotating savings and credit association (ROSCA) is considered as one of the most prevalent forms of informal financial institutions in developing countries. Similar notions of ROSCAs are shared by different places in the world. The basic principle is that a group of people gathers together and, in each meeting, everybody contributes to a common money pot, which is given to only one member of the group each time until every member has obtained the money pot once.

As observed from real cases and related literatures, one important characteristic of a ROSCA is that it originally emerged among people who lived in isolated areas and who had limited access to formal financial institutions. Thus, when someone needed an item of lumpy consumption (such as for house construction, a wedding, treatment for disease, or other emergency use), (s)he can only rely on personal borrowing from relatives or friends. Kovsted and Lyk-Jensen (1999) described the origin of a ROSCA as a "private borrowing-lending club", which highlights its mutual help function in its developing history. This history may explain the mechanism formation process of a ROSCA when the mutual-aid action frequently happens among the same group of people.

Another key feature is that contributions to the money pot are voluntary. If all participants contribute to the fund after they have received their payment, the ROSCA is managed successfully (Koike et al. 2010). In this sense, it requires strict cooperation; otherwise, the system collapses, even if only one member of the group defaults.

At the same time, abundant studies have discussed the motivations of joining ROSCA empirically. We group these motivations into two types: *direct benefit* and *indirect benefit*. *Direct benefit* contains: (1) the early pot motive (Besley et al. 1993; Anderson and Kilduff 2009; Bisrat et al. 2012) that allows participants to purchase durable goods or invest in a family business earlier; (2) extra interest income compensating participants who obtain the money pot relatively late (Sandsør 2010); (3) insurance motive (Calomiris and Rajaraman 1998; Klonner 2003); and (4) the commitment device model (Ashraf et al. 2006; Gugerty 2007). As for the *indirect benefit*, we surmise it is related to indirect reciprocity. Ambec and Treich (2007) describe it as a kind of social pressure, which, we theorize, becomes a social mechanism that insures a person will be helped when he or she needs it; similarly, people do participate in a ROSCA through indirect reciprocity.

As a widely utilized way of cooperation, a ROSCA shows various schemes with huge temporal and geographical disparities. However, we observed that, normally, only one type of ROSCA is dominant in each region, which stimulates our research about how these rule disparities occur and, most importantly, whether people's choices among various rules of a ROSCA are affected by different reciprocity levels in these regions. Yet, the effect from reciprocity level on the rule preference of a society is rarely reported. A ROSCA supplies us with an empirical case to study the evolution of cooperation in terms of how different schemes emerge and its relationship with reciprocity. The above incentives, seemingly mixed up to some extent, give us sufficient implications to discuss the potential factors behind the mechanism.

ROSCA has been practiced as a useful informal financial institution for several hundred years all over the world. The purpose of this paper is to discuss the relationship between rule preference and reciprocity level when people choose among different ROSCA types. In this paper, we build an evolutionary imitation game regarding the participants' choice of ROSCA rules. Previous research has barely demonstrated that the evolution of cooperation can be explained as outcomes of microscopic interactions between altruistic but also self-motivated individuals. The original contribution of this paper is considered as modeling people's mutual aid behaviour and showing how different rules are chosen by the society.

Therefore, in this paper we focus on the evolution of cooperation by taking ROSCAs in China as a case study. A social learning model on how reciprocity level affects people's rule preference is proposed. We assume that individuals change their choice through interactions with others following an imitation rule. Finally, we want to provide new insights to analyse the factors underlying the rule preferences among different schemes of ROSCAs, and show its evolutionary direction to contribute to the practical utilization of evolutionary cooperation theories.

This paper is organized as follows. Section 2 discusses ROSCA schemes and defines the types of random ROSCAs. In Sect. 3, we show the four types of ROSCAs practiced in China and calculate the function of each type. In Sect. 4, we build up random pairwise imitation game models, in which participants interact with. The money discounting factor, δ, is adopted in this model. We identify the reciprocal factor, θ, as the weight of caring about other people's utility. Section 5 describes the experimental procedure and discusses the results. Section 6 concludes the paper.

2 ROSCA Schemes

The considerable literature on ROSCAs reveals many variations about how they work in practice (Nguyen and Tanaka 2010; Bouman 1995; Geertz 1962). Their rules and mechanisms, in terms of determining the order of rotation, have been widely discussed and have been used as a classification criterion in the ROSCA literature, which include the *random ROSCA* and the *bidding ROSCA*. They primarily differ by the ordering method of receiving the pot among members. In this paper, we follow Besley's (1995) and Kovsted and Lyk-Jensen's (1999) classifications and categorize them as the following three types:

a. The random ROSCA: The order of obtaining the money pot is decided by lottery before the whole ROSCA starting or before each meeting.
b. The bidding ROSCA: The money pot is given to the person who gives the highest bid in that meeting.
c. The mixed ROSCA: The order of allocating the pool is decided by some predefined criterion, including the social influence of the leader(s) and/or members, degree of emergency, social network, and reputation.

In the field, the random ROSCA is more common and the rules are more flexible; sometimes it appears with mixed types (in which the partial orders of obtaining the money pot can be negotiated, especially when someone is in urgent need). In this paper, we focus on the random ROSCA. Based on the literature and fieldwork, under the category of random ROSCA, we have observed rule disparities, which can be categorized into several types. We define the details of each type as the following.

The common rule for all types in a random ROSCA is that the order of obtaining the money pot is decided by lotteries before the ROSCA begins. Then we identify the distinguishing characteristics of random ROSCA types by the following rules:

(1) Payment can be consistent or not in every meeting for each order. The payments in each meeting accumulate into the money pot given to one of the members, which may result in disparities in total payment and obtainment for each participant among different ROSCA types.

(2) We have mentioned that the order of obtaining the money pot is decided by a one-off lottery beforehand. However, due to the specialties of ROSCA, including self-organizing, flexibility, and reciprocity, partial orders can be negotiated among the participants. For example, a group leader is responsible to organize the meeting and other issues. Thus, in some types, the leader can choose the order of obtaining the money pot. As for the reciprocal reason, an individual in urgent need can be arranged in the earlier order. In Sect. 3, we will demonstrate the details of different ROSCA types in China following the above-mentioned rules.

3   ROSCA Type in China

A ROSCA usually happens among relatively homogenous people who live close to each other and have similar consuming abilities or needs, with limited outside financial resources.

Let $X = \{1, 2, ..., m, ..., M\}$ denote a society with M individuals. Each individual m has the property θ_m, in which θ_m denotes the reciprocal factor for individual m. Let $s_i \in I \times F$ denote a ROSCA group information set, whereas s_i varies depending on the ROSCA type i and detailed parameter values in F. Let $F = \{\varphi_i\}_{i \in I}$ denote the set of ROSCA group alternatives, where $I = \{1, 2, 3, ...\}$ is the set of ROSCA type, which indicates different rules of different types, and $\varphi_i \subseteq \Theta = \{w, \alpha, d, \eta\}$ denote the set of ROSCA rule parameters, which will be explained below in this section.

As mentioned above, ROSCAs have appeared all over the world for a long time; therefore, it is not possible to enumerate all types of ROSCA in the field. In this paper, four main types of ROSCA in China, including the calculating equations, are exhibited in 3.1 to 3.4 in this section. Based on these types, how evolution and coexistence occur in each model will be discussed.

Briefly, in China, ROSCAs are known as *Hehui* and act as civilian financial organizations; these were established as early as the Tang Dynasty (618–907) Wang (1930). In the latter part of the Qing Dynasty and the Republic of China (1912–1948), *Hehui*'s development reached its peak. With the establishment of the People's Republic of China in 1949 and the launch of the planned economy, *Hehui* almost stagnated. Since the reforms of the 1970s, the commodity economy has developed and *Hehui* revived again spontaneously.

The main difference among these four types of ROSCA is their respective way to decide the payment of each player at each meeting. Here, let $a_i(k,n)$ denote the payment at the kth meeting for the player who obtains the money pot at n^{th} order in one ROSCA group that belongs to type i. Therefore, for each type of ROSCA below, the formula to calculate $a_i(k,n)$ is shown respectively. We assume that N is the number of people participating in this group. One of the characteristics of a ROSCA is to aid people in urgent need of money by placing them in an earlier order (normally, first); thus, a participant's savings target will be decided beforehand. Here, let w denote the target money pot under one ROSCA rule, which denotes how much money is accumulated and given for the first participant. Thus, w is decided by the members before each ROSCA begins.

Suppose all participants have different time-related preferences and are aware of it. At each meeting of one type, all the players contribute a pre-decided amount to form a money pot and one of the players takes all the funds collected at this meeting; the money pot thus becomes empty. At the next meeting, players refill the money pot in the same way and another player takes all the money back. After N meetings (i.e., all the players have taken the money once in turn), this round of ROSCA stops.

To show the rules more intuitively, real cases from the field are given in each type to explain how much a participant needs to pay at every ROSCA meeting. All the schemes are budget balanced.

3.1   Type 1: Suojin ROSCA

The first type is the Suojin ROSCA and is characterized by φ_1, including number of participants N, money pot w, and a parameter, α ($0 \leq \alpha \leq 1$), indicating the ratio between the payment of the player who receives the amount at the last meeting and the payment of the player who does at the first meeting. The payment of each member for each time keeps decreasing until the member obtains the pot; then, the money changes into a fixed number (in the case of Table 1, the fixed number is 10,556 yuan). The Suojin ROSCA has been popular in the South East area of China since the Qing Dynasty when informal financial institutions and small family businesses began and became prosperous. The rule of calculating the amount one member must pay at each meeting is much more complex than the traditional method described below and demands a higher education from the participants Wang (1930). The payment at each meeting of each person $a_1(k,n)$ is calculated based on Eq. (1):

$$a_1(k,n) = \begin{cases} \frac{w-c(k-1)}{N-k+1}, & 1 \leq k \leq n \\ c, & n < k \leq N \end{cases} \quad (1)$$

where

$$c = \frac{w\left(1 - \frac{\alpha}{N}\right)}{N - 1}$$

In Table 1, N = 10, w = 100,000, and α = 0.5.

Table 1. Flow of money in Suojin ROSCA

Meeting (k)/order (n)	1	2	3	4	5	6	7	8	9	10
1st meeting	10000	10000	10000	10000	10000	10000	10000	10000	10000	10000
2nd	10556	9938	9938	9938	9938	9938	9938	9938	9938	9938
3rd	10556	10556	9861	9861	9861	9861	9861	9861	9861	9861
4th	10556	10556	10556	9762	9762	9762	9762	9762	9762	9762
5th	10556	10556	10556	10556	9630	9630	9630	9630	9630	9630
6th	10556	10556	10556	10556	10556	9444	9444	9444	9444	9444
7th	10556	10556	10556	10556	10556	10556	9167	9167	9167	9167
8th	10556	10556	10556	10556	10556	10556	10556	8704	8704	8704
9th	10556	10556	10556	10556	10556	10556	10556	10556	7778	7778
10th	10556	10556	10556	10556	10556	10556	10556	10556	10556	5000
Total payment	105004	104386	103691	102897	101971	100859	99470	97618	94840	89284
Total income	100000	100000	100000	100000	100000	100000	100000	100000	100000	100000
Net income	-5004	-4386	-3691	-2897	-1971	-859	530	2382	5160	10716

3.2   Type 2: Shensuo ROSCA

The second type, Shensuo ROSCA, is characterized by φ_2, including number of N participants, money pot w, and a parameter, d, which stands for a payment gap between each meeting for the first-order player. As the leader's premium, he obtains the money pot at the first order without paying extra interest. The payment at each meeting of each person $a_2(k, n)$ is calculated based on Eq. (2):

$$a_2(k, n) = \begin{cases} 0, & k = n \\ h - d(n - 2), & k \neq n \end{cases} \tag{2}$$

where

$$h = \frac{w}{N - 1} + \frac{(N - 2)d}{2}$$

In Table 2, $N = 10$, $w = 100{,}000$, and $d = 100$.

Table 2. Flow of money in Shensuo ROSCA

Meeting (k)/order (n)	1	2	3	4	5	6	7	8	9	10
1st meeting	0	11511	11411	11311	11211	11111	11011	10911	10811	10711
2nd	11511	0	11411	11311	11211	11111	11011	10911	10811	10711
3rd	11411	11511	0	11311	11211	11111	11011	10911	10811	10711
4th	11311	11511	11411	0	11211	11111	11011	10911	10811	10711
5th	11211	11511	11411	11311	0	11111	11011	10911	10811	10711
6th	11111	11511	11411	11311	11211	0	11011	10911	10811	10711
7th	11011	11511	11411	11311	11211	11111	0	10911	10811	10711
8th	10911	11511	11411	11311	11211	11111	11011	0	10811	10711
9th	10811	11511	11411	11311	11211	11111	11011	10911	0	10711
10th	10711	11511	11411	11311	11211	11111	11011	10911	10811	0
Total payment	100000	103600	102700	101800	100900	100000	99100	98200	97300	96400
Total income	100000	100000	100000	100000	100000	100000	100000	100000	100000	100000
Net income	0	-3600	-2700	-1800	-900	0	900	1800	2700	3600

3.3   Type 3: Duiji ROSCA

The third type is Duiji ROSCA and characterized by φ_3, including number of participants N, money pot w, and a parameter, η (>1), which stands for an increased payment ratio after each player takes the money. This type is popular due to its ease of handling and extra benefits that are compensated to the later recipients. At the same time, the shortcoming is obvious: the earlier recipients pay too much interest. The payment at each meeting of each person $a_3(k, n)$ is calculated based on Eq. (3):

$$a_3(k, n) = \begin{cases} \frac{w}{N}, 1 \leq k \leq n \\ \frac{\eta w}{N}, n < k \leq N \end{cases} \tag{3}$$

Table 3. Flow of money in Duiji ROSCA

Meeting (k)/order (n)	1	2	3	4	5	6	7	8	9	10
1st meeting	10000	10000	10000	10000	10000	10000	10000	10000	10000	10000
2nd	11000	10000	10000	10000	10000	10000	10000	10000	10000	10000
3rd	11000	11000	10000	10000	10000	10000	10000	10000	10000	10000
4th	11000	11000	11000	10000	10000	10000	10000	10000	10000	10000
5th	11000	11000	11000	11000	10000	10000	10000	10000	10000	10000
6th	11000	11000	11000	11000	11000	10000	10000	10000	10000	10000
7th	11000	11000	11000	11000	11000	11000	10000	10000	10000	10000
8th	11000	11000	11000	11000	11000	11000	11000	10000	10000	10000
9th	11000	11000	11000	11000	11000	11000	11000	11000	10000	10000
10th	11000	11000	11000	11000	11000	11000	11000	11000	11000	10000
Total payment	109000	108000	107000	106000	105000	104000	103000	102000	101000	100000
Total income	100000	101000	102000	103000	104000	105000	106000	107000	108000	109000
Net income	-9000	-7000	-5000	-3000	-1000	1000	3000	5000	7000	9000

In Table 3, $N = 10$, $w = 100,000$, and $\eta = 1.1$.

3.4   Type 4: Traditional ROSCA

Traditional ROSCA was a type that used to be popular in China. According to our interviews, the basic principle of this traditional type is that when there is someone in need of money (for example, in an emergency or for a big event), relatives and neighbours gather together and contribute to a fund pot and one of them receives it in rotation; thus, in this type, no extra interest is required. In terms of motivation for the participants without emergency needs, their answer is, basically, "I help other people this time, and they will also help me next time." Therefore, even without the extra benefit, people were still willing to join in. The rule of this type is the simplest one compared to the other three types. That is, all the members pay the same amount of money in every meeting; this is characterized by φ_4, including number of participants N and money pot w. The payment at each meeting of each person $a_4(k, n)$ is calculated based on Eq. (4):

$$a_4(k, n) = \frac{w}{N} \tag{4}$$

Table 4. Flow of money in traditional ROSCA

Meeting (k)/order (n)	1	2	3	4	5	6	7	8	9	10
1st meeting	10000	10000	10000	10000	10000	10000	10000	10000	10000	10000
2nd	10000	10000	10000	10000	10000	10000	10000	10000	10000	10000
3rd	10000	10000	10000	10000	10000	10000	10000	10000	10000	10000
4th	10000	10000	10000	10000	10000	10000	10000	10000	10000	10000
5th	10000	10000	10000	10000	10000	10000	10000	10000	10000	10000
6th	10000	10000	10000	10000	10000	10000	10000	10000	10000	10000
7th	10000	10000	10000	10000	10000	10000	10000	10000	10000	10000
8th	10000	10000	10000	10000	10000	10000	10000	10000	10000	10000
9th	10000	10000	10000	10000	10000	10000	10000	10000	10000	10000
10th	10000	10000	10000	10000	10000	10000	10000	10000	10000	10000
Total payment	100000	100000	100000	100000	100000	100000	100000	100000	100000	100000
Total Income	100000	100000	100000	100000	100000	100000	100000	100000	100000	100000
Net Income	0	0	0	0	0	0	0	0	0	0

In Table 4, $N = 10$ and $w = 100,000$.

4   Model

The purpose of this paper is to analyze a participant's rule preference among different random ROSCA types. The model developed here closely describes the different observed rules of a ROSCA. We consider that those different types appearing in a ROSCA reflect the different levels of economic development, pecuniary needs, reciprocal levels, and so on. These varieties show the high flexibility and rational responses of a random ROSCA, which allows maximizing the expected utility of its members.

4.1   Pecuniary Payoff from a ROSCA

In one round of ROSCA, the revenue of the player who receives the money pot at the nth order, hereby denoted as $C_n(s_i)$, depends on the payment that each player makes at each meeting, which can be calculated as the following Eq. (5):

$$C_n(s_i) = \sum_{n'=1}^{N} a_i(n, n')$$

(5)

By definition, for ROSCA types 1, 2, and 4, $C_n(s_i)$ is constantly equal to the money pot w: i.e., $C_n(s_1) = C_n(s_2) = C_n(s_4) = w$ for $^\forall n$. For type 3, the revenue that each participant receives depends on the order, which is given by $C_n(s_3) = \frac{nw + (N-k)\eta w}{N}$. Thus, for simplicity of notation, $C_n(s_i)$ is used to denote the income for each participant.

Let $u(n|s_i))$ denote a player's net pecuniary payoff who chooses group s_i. The pecuniary payoff is affected by the discount factor of the society δ, the ROSCA group he participated in, and his order of taking the money. This relationship is described in the following Eq. (6):

$$u(n|s_i) = C_n(s_i) \cdot \delta^{n-1} - \sum_{k=1}^{N} a_i(k,n) \cdot \delta^{k-1} \qquad (6)$$

4.2 Reciprocity

As an outstanding example of cooperation in human history, reciprocal behaviour in a ROSCA is widely known and discussed, and is one of the main incentives encouraging people to participate. Therefore, it is straightforward to assume that rule preference of a society with a higher level of reciprocity is different from a society with a lower one.

Direct reciprocity relies on repeated encounters between the same two individuals, but often the interactions among humans are asymmetric. Helping someone by participating in a ROCSA establishes a good reputation, which may help in getting assistance from others in the future. The reputation-accumulating mechanism in a ROSCA is called indirect reciprocity and captured in the principle: "I help you this time, and when I need help, someone else will help me back" (Nowak and Sigmund 2005).

Normally, a ROSCA is established for a person in urgent need who wants to obtain the money pot in the first order. In this sense, members in a ROSCA do not only care for their own monetary benefit, but also value the utility of another certain person who needs help. The basic postulate is that individuals' concern toward others can be characterized by reciprocal utility functions. For the sake of simplicity, let us consider a person who obtains the money pot at the 1st order is solely in urgent need of money. In this sense, the 1st order of obtaining the money pot has been decided in advance. For the others, their order is decided randomly by a lottery. Therefore, the expectation of monetary benefit for those participants not receiving the money pot at the 1st order in this ROSCA (denoted with the suffix _1) is calculated as Eq. (7):

$$E_{-1}(s_i) = \frac{1}{N-1} \sum_{n \neq 1} u(n|s_i) \qquad (7)$$

The reciprocity characteristic of players is considered herein as one of the incentives to participate in a ROSCA. The expected utility of participant m (denoted as v) is thus affected by players' reciprocal parameter θ_m and utility of the first order obtainment player denoted as $u(1|s_i)$, and as expressed in Eq. (8). Note that this expected utility no longer depends on the order in which the player m receives the money pot.

$$v_{-1}(m|s_i) = (1 - \theta_m)E_{-1}(s_i) + \theta_m u(1|s_i) \qquad (8)$$

Also, note that $\theta_m u(1|s_i)$ shows that how much player m cares about the first recipient's benefit in his group with the weight of θ_m.

4.3 Imitation Game in a Random Encounter

Assuming each individual decides which ROSCA to join based on his/her reciprocal utility in Eq. (8), the evolutionary dynamism of human cooperation produced through ROSCA is very complex and of a nonlinear nature. In the real human society, such behaviours also involve trial and error to find a more desirable ROSCA rule. As an important part of game theory, evolutionary imitation game theory tries to explain how a new behaviour or a new rule is diffused among the whole society by assuming that people adopt new things when they encounter others who have already adopted them. In this paper, we follow this theory and an imitation game is proposed in this section as an attempt to imagine the trajectory of rule preference changes, which allows participants to learn about new ROSCA types by encountering and learning from others. The basic assumption is that when adopting a new ROSCA rule, each individual makes a rational choice of a ROSCA type that maximizes his or her payoff as compared to the last choice according to his/her degree of reciprocity. We set the imitation rule as to when and how each participant changes his/her actions in interacting with other participants as the following.

At the initial round, an individual m randomly joins a ROSCA group s_p of type p. He receives a revision opportunity when he finishes one round of ROSCA and randomly encounters another member in the society who does not belong to the s_p group. He observes the encounter's ROSCA type information set (for example, group s_q) and its expected payoff. Then, this participant imitates the actions of someone he met and considers joining one of their ROSCA groups from the next round if the expected utility of the s_q group is greater than that of s_p group according to his personal information θ_m.

Ideally, a ROSCA group with a higher utility leads to a higher probability that it is chosen. However, as previously mentioned in the first section, a ROSCA is a cooperative behaviour in which the number of participants in one group is required and predetermined (e.g., a ROSCA with a group size of 5 cannot be operated by only 3 players). Hence, in our simulation game, a player can actually join a group when the exact number of people are willing to sign in the same group in the next round; otherwise, players who fail to gather a group have to wait until next round.

5   Numerical Simulation and Discussion

In this section, numerical simulation results based on the above model are demonstrated by programming on Matlab 2014. Following the aforementioned updating principles, this paper attempts to understand how different distributions of reciprocity influence participants' preferences on the four types of ROSCA. Suppose there are 36,000 players in the society and they do not have any access to financial institutions except for participating in a ROSCA, and once they participate in a ROSCA group, no one will default.

In the initial round, all players in the society are allocated with the same portion into the four ROSCA types (9,000 players in each). For each ROSCA type, 10 different money pot values are considered, ranging from 10,000 to 100,000, with an interval of 10,000. Similarly, group capacity varies from 5 to 40 with intervals of 5. As a result, 80 subtypes are generated through combining different group sizes and money pot values. The discounting factor δ is set to be 0.98.

The random encounter happens when the first round of a ROSCA is finished. For each player, he encounters one other player in the society and obtains information about a new ROSCA type only from the newly encountered person. The player then decides whether he changes group or not, and another round of a ROSCA starts. At each round, there may exist some players failing to join any group and they keep waiting until the next round. This process continues until the portion of each type converges to a stable value. The final ratios of the four main types are cumulated from the total number of each subtype.

Figure 1 illustrates a typical simulation outcome in a *selfish* society where reciprocity parameter θ distributes uniformly on (0.2, 0.3). We notice that the type 1 ROSCA dominates in the society with more than 70% of the population. In contrast, in an *altruistic* society where reciprocity parameter θ distributes uniformly on (0.8, 0.9), the type 4 is the most popular ROSCA with 60% of the population choosing it, whereas type 3 is the second most popular ROSCA in the society as Fig. 2 shows. Due to the simulation parameter settings presented in this paper, four types of ROSCAs may coexist, but finally there is often a popular choice in the whole society with the highest probability, which is also a widely observed phenomenon in the field.

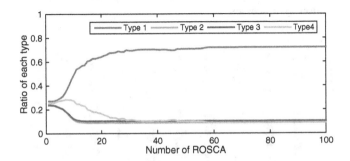

Fig. 1. Selfish society where θ is uniformly distributed on (0.2, 0.3)

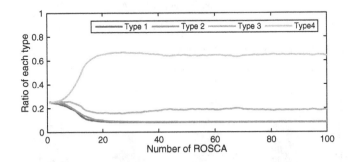

Fig. 2. Altruistic society where θ is uniformly distributed on (0.8, 0.9)

To show a more detailed picture of the relationship between reciprocity and rule preferences in a ROSCA game, the value of θ is divided into 10 partitions with an interval of 0.1 from 0 to 1 as shown in Figs. 3 and 4 partitions with an interval of 0.7 as Fig. 4 shows. Each portion represents the mean converged share of its corresponding ROSCA type (out of $100 \times 1,000$ trials) in which the reciprocal parameter θ is uniformly distributed across each interval. Within a 0.1 interval, the results in Fig. 3 show a homogeneous society where all players share the similar reciprocal value, while a more heterogeneous society is shown in Fig. 4 and some different results are observed.

In a more homogeneous society demonstrated in Fig. 3, the share of type 1 reaches the peak where the value of θ in the society is distributed at both a (0.2–0.3) and (0.3–0.4) reciprocity scale. The portion of type 2 is comparatively bigger in a selfish (0–0.2) and altruistic (0.8–1) society, while type 3 attracts more people only in a selfish society (especially around 0 to 0.2 reciprocity scale) and loses its advantage when the society becomes more reciprocal. Type 4 gradually attracts more people and becomes the dominant type when θ exceeds 0.5. Similar trends are observed in a more heterogeneous society in Fig. 4. However, types 1 and 4 are always the first or second dominant type in each θ interval, but neither holds more than a 50% portion. Those results help to explain why, in a certain society, there always exists a dominant type of cooperation. At the same time, it is worth noting that, while different reciprocity settings produce different equilibria, under any setting no rules completely disappeared during the simulated evolutionary processes.

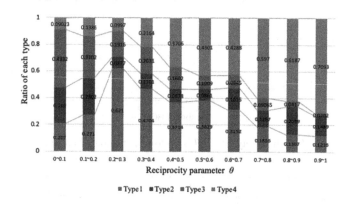

Fig. 3. Ratio of four ROSCA types on different θ partitions (0.1 interval)

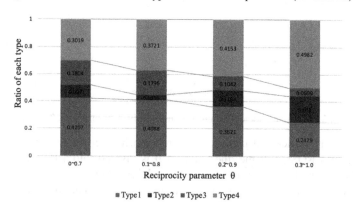

Fig. 4. Ratio of four ROSCA types on different θ partitions (0.7 interval)

Some empirical investigations reveal that social rules keep evolving and people share rules by learning from each other. In the field, disparities in the rules are observed not only in China, but also in regions where ROSCAs are popular. Similarly, some traditional rules, as type 4 demonstrated in this study, are still widely utilized. However, new types that demand extra interest have also emerged, and are considered as evidence that participants are realizing that the value of time and money is decreasing. Therefore, even when the original incentive to participate in a ROSCA is still to help other people, the helping side also does not want to lose their monetary benefit. Findings from both the field and our simulation in this paper supply us with concrete support for arguing that participants' rule preference of a ROSCA is related to the level of reciprocity and net benefit by joining a ROSCA group. While no prior empirical studies have so far succeeded in specifying a mechanism as to how members' interactions promote cooperation, our simulations do explain how this dynamism can be produced under various conditions.

6 Conclusion

Four types of ROSCAs in China were modeled with the consideration of individual investment ability and the discounting parameter. Although these equations and parameters are largely specific to the Chinese context, rules can be extended into different countries and conclusions can be made following this model. The reciprocity function was proposed and a numerical analysis was conducted to confirm whether rule preferences of participation in a ROSCA correlate with reciprocity levels. We argue that the evolution of ROSCA is an indicative exemplar in understanding the evolution of cooperation. The following conclusions are drawn:

(1) In different societal settings, there is a dominant ROSCA type; but other types may coexist. These results can explain the current ROSCA situation in China from some aspects. The four types of ROSCA have a long history in China; however, even with the passage of time, these types still coexist in different areas (societies).

(2) Different ROSCA rules directly influence participants' monetary benefit. Simulation results show that the value of reciprocity has a significant effect on people's rule preference.

(3) In social transition processes, it is crucial to study individual rule preference, which may influence the evolutionary direction of social norms. Our results suggest that financial institutions require some knowledge about the private information and preferences of potential participants; such knowledge can provide insights into the theory of collective rule choice that triggers the evolution of cooperation.

Note that, apart from the factors discussed here, there are still many other factors which can also affect people's rule preference. Simulation and mathematical models are effective for exploring possible dynamisms of social interactions when there is a large number of people involved. Future work would include providing feedback on experiments and empirical research by presenting this causal relationship. While the specific results from this study may not extend directly to all social conditions, they help to better understand the reciprocity level in a society, which significantly influences people's rule preferences.

References

Trivers, R.L.: The evolution of reciprocal altruism. Q. Rev. Biol. **46**(1), 35–57 (1971)

Axelrod, R., Hamilton, W.D.: The evolution of cooperation. Science **211**(4489), 1390–1396 (1981)

Nowak, M.A.: Five rules for the evolution of cooperation. Science **314**(5805), 1560–1563 (2006)

Nowak, M.A., Sigmund, K.: Evolution of indirect reciprocity. Nature **437**(7063), 1291 (2005)

Koike, S., Nakamaru, M., Tsujimoto, M.: Evolution of cooperation in rotating indivisible goods game. J. Theor. Biol. **264**(1), 143–153 (2010)

Besley, T., Coate, S., Loury, G.: The economics of rotating savings and credit associations. Am. Econ. Rev. **83**, 792–810 (1993)

Anderson, C., Kilduff, G.J.: The pursuit of status in social groups. Curr. Dir. Psychol. Sci. **18**(5), 295–298 (2009)

Bisrat, A., Kostas, K., Feng, L.: Are there financial benefits to join RoSCAs? Empirical evidence from equb in Ethiopia. Procedia Econ. Finan. **1**, 229–238 (2012)

Sandsør, A.M.J.: The rotating savings and credit association: an economic, social and cultural institution (2010)

Calomiris, C.W., Rajaraman, I.: The role of ROSCAs: lumpy durables or event insurance? J. Dev. Econ. **56**(1), 207–216 (1998)

Klonner, S.: Rotating savings and credit associations when participants are risk averse. Int. Econ. Rev. **44**(3), 979–1005 (2003)

Ashraf, N., Bohnet, I., Piankov, N.: Decomposing trust and trustworthiness. Exp. Econ. **9**(3), 193–208 (2006)

Gugerty, M.K.: You can't save alone: commitment in rotating savings and credit associations in Kenya. Econ. Dev. Cult. Change **55**(2), 251–282 (2007)

Ambec, S., Treich, N.: Roscas as financial agreements to cope with self-control problems. J. Dev. Econ. **82**(1), 120–137 (2007)

Tanaka, T., Camerer, C.F., Nguyen, Q.: Risk and time preferences: linking experimental and household survey data from Vietnam. Am. Econ. Rev. **100**(1), 557–571 (2010)

Bouman, F.J.A.: Rotating and accumulating savings and credit associations: a development perspective. World Dev. **23**(3), 371–384 (1995)

Geertz, C.: The rotating credit association: a "middle rung" in development. Econ. Dev. Cult. Change **10**(3), 241–263 (1962)

Kovsted, J., Lyk-Jensen, P.: Rotating savings and credit associations: the choice between random and bidding allocation of funds. J. Dev. Econ. **60**(1), 143–172 (1999)

Wang: The history of Chinese Hehui. Zhongguo Hezuo Xueshe (1930)

Tanaka, T., Nguyen, Q.: ROSCA as a saving commitment device for sophisticated hyperbolic discounters: field experiment from Vietnam. Working Paper, Arizona State University (2010)

Besley, T.: Savings, credit and insurance. In: Behrman, J, Srinivasan, T.N. (eds.) Handbook of Development Economics, pp. 2123–2207 (1995)

Modeling the Conflict Within Group Decision Making: A Comparison Between Methods that Require and Do Not Require the Use of Preference Aggregation Techniques

Alexandre Bevilacqua Leoneti$^{(\boxtimes)}$ ⓘ and Vanessa Coimbra Ziotti ⓘ

Research Group in Decision Sciences, School of Economics,
Business Administration and Accounting, University of São Paulo,
Ribeirão Preto, Brazil
ableoneti@usp.br

Abstract. This paper compares MCDM methods that require and do not require the use of aggregation techniques for their application within group decision making process and study the associated concern regarding the possibility of underestimation of conflicts. An experiment was adapted from previous research and three MCDM methods (TOPSIS, TOPSIS for group, a method based on game theory) were empirically compared based on the performance of these methods to predict correctly the agreement reached by the group. A criterion was established to be used as a performance indicator, which was the counting of the matches between the agreements of the group (if reached) with the solution of the MCDM method. It was very strong the difference in matches between the methods that require and do not require the use of preference aggregation techniques, which seems to corroborate the affirmation that modeling the conflict by using a more adequate methodology for dealing with conflictive scenarios provides best efficiency in predicting group decision making outcomes.

Keywords: Multi-criteria Decision-Making · TOPSIS · Game theory ·
Nash equilibrium · Negotiation

1 Introduction

The use of Multicriteria Decision Making Methods (MCDM) is nowadays recurrent in both academic and organizational environment, where group decision making is an ordinary assignment. A regular practice for using MCDM methods within group decision making process is the use of aggregation techniques. These techniques intend to amalgamate the preferences of different decision makers into a unit that would represent the group to make possible the utilization of MCDM methods, which are predominantly designed for individual use. Among the most used aggregation techniques are the weighted arithmetic mean and the weighted geometric mean that can be employed externally or internally to the MCDM methods [1]. Notwithstanding the disseminated use of these techniques, little attention is dedicated to the fact that decision makers

© Springer Nature Switzerland AG 2019
D. C. Morais et al. (Eds.): GDN 2019, LNBIP 351, pp. 57–64, 2019.
https://doi.org/10.1007/978-3-030-21711-2_5

preference's might be diametrically different and that the aggregation of preferences may dissimulate potential conflicts [2, 3].

One of most employed MCDM method is the Technique for Order Performance by Similarity to Ideal Solution (TOPSIS) [2]. Most of applications of TOPSIS in group decision making process is by the use of external aggregation using either weighted arithmetic mean or the weighted geometric mean for amalgamate the weighting vectors of decision makers involved. The application of TOPSIS within group decision making can be summarized in seven steps, which are: (i) decision matrix creation; (ii) decision matrix standardization; (iii) elicitation of the weighting vector of each decision maker and aggregation of these vectors for the purpose of weighting the standardized decision matrix; (iv) determination of the worst alternative and the best alternative; (v) calculation of the distance between each alternative to the best and worst alternative; (vi) calculation of the similarities index; and (vii) ranking the alternatives according to the calculated similarities. The procedure in step three is considered an example where the aggregation process occurs externally to the core of the calculations of the method.

According to [1], to achieve a better efficiency of MCDM methods within group decision making process it would be favorable the internalization of the aggregation process into the core of the method. Toward a more suitable version of TOPSIS for group decision making, Shih et al. [1] proposed to weight distances instead of weighting the decision matrix, as in the original version of the TOPSIS method. The steps for applying the extended version of TOPSIS for group decision making are slightly different from the original version and can be summarized in the following seven steps: (i) decision matrix creation for each decision maker; (ii) decision matrix standardization for each decision maker; (iii) determination of the worst alternative and the best alternative for each decision maker; (iv) calculation of the weighted distance between each alternative to the best and worst alternative by using the weighting vector for each decision maker; (v) aggregation of the distances between each alternative to the best and worst alternative for each decision maker; (vi) calculation of the similarities index to the group; and (vii) ranking the alternatives according to the calculated similarities. It should be noted that, tough internally, the use of the aggregation procedure is still present.

Recalling that an associated concern regarding the use of aggregation techniques in the application of MCDM methods within group decision making process is the one related to the underestimation of conflicts, we stress that game theory has also been widely applied in both academic and organizational environment for modeling and solving group decision making problems specially in the presence of conflictive scenarios and without the need of using aggregation techniques. Towards a generalized MCDM method for modeling and solving group decision making as games, Leoneti [3] proposed a utility function based on pairwise comparison of alternatives that allows to solve multi-criteria and multi-agent problems without the use of the aggregation procedure. The steps for applying the method can be summarized into the following seven steps: (i) decision matrix creation for each decision maker; (ii) decision matrix standardization for each decision maker; (iii) elicitation of the weighting vector of each decision maker for the purpose of weighting the standardized decision matrix of each decision maker; (iv) application of the utility function for translating each standardized and weighted decision matrix into payoff matrixes for each decision maker; (v) application of an equilibrium solution concept for solving the game; (vi) application of a social welfare ordering function when more than one equilibrium is found; and (vii) selection of the equilibrium according to the social welfare ordering applied. According to [3], the main advantage of using this utility function for modeling group decision making problems as games is the use of game theory approach to circumvent the limitation of the aggregation techniques that are usually employed in group MCDM applications.

In this sense, we propose to empirically compare the aforementioned methods in order to verify whether the performance of the methods that do not require the use of aggregation techniques overcome the ones that do require them in terms of predicting group decision making outcome.

2   Materials and Methods

The first stage of the research was the selection of an experiment to simulate group decision making to evaluate the performance of three methods in terms of predicting correctly group decision making outcome, henceforth called agreement. The chosen methods were: (i) original TOPSIS by [1]; (ii) extension of TOPSIS for group decision making by [2]; and (iii) utility function by [3] for modeling multi-criteria and multi-agent decision making problems as games. The TOPSIS method was chosen since it is one of most used MCDM method nowadays and the utility function was conveniently chosen since it is an example of MCDM method based on game theory that does not require the use of aggregation techniques.

2.1   The Experiment

The experiment was adapted from previous research of Leoneti and de Sessa [4]. The decision problem was concerned to the choice of a travel destination to be held in group. In order to provide information regarding five possible travel destinations, Leoneti and de Sessa [4] created a decision matrix with eight criteria (hotel evaluation, travel time in hours, length of stay in nights, cost in reais, shopping facilities, cultural attractions, natural landscapes, and safety) based on five real travel packages offered by Brazilian travel companies. Here, the original decision matrix was adapted to the one shown in Fig. 1, replacing the criterion "cost" by "exchange rate in relation to reais" and "safety" by "infrastructure". A number of 125 undergraduate students from the School of Economics, Business Administration and Accounting at the University of São Paulo in Ribeirão Preto participated in 25 simulations, in which five participants were selected in each of them. The presentation of the simulation was as follows: "In order to attract and retain customers, a travel company held a promotion. A group of people was randomly selected to travel with all expenses paid by the company. The promotion conditions are: winners must travel together and the agency will cover hotel expenses (including breakfast) and transfers. Congratulations, you were one of the lucky ones! Since all winners have at least 12 days of vacation, you must negotiate with the other winners to decide the travel destination". Following, the participants were invited to analyze the decision matrix individually and to rank the criteria ascending. The participants were also invited to rank the alternatives for negotiating the travel destination in a negotiation round that occurred after the individual analysis. Finally, the group communicate the agreement (if reached) to the authors of the present research.

Alternatives	Criteria							
	Hotel evaluation	Travel time (hours)	Length of stay (nights)	Exchange rate	Shopping	Cultural attractions	Nature	Infrastructure
A	2.5	8	4	0.9	5	3	7	8
B	3.5	2.5	6	3.1	9	7	3	6
C	3	4	7	4.7	4	5	9	7.5
D	5	13	5	3.3	6	9	6	7
E	4	16	8	1.1	3	8	5	5

Fig. 1. Decision matrix

2.2 Elicitation of the Weighting Vectors

The ranking of criteria given by each decision maker were translated into weighting vectors by using the Ranking Order Centroid – ROC method [5]. The ROC method is a type of rank-order method, which applies a transformation of ranks into ratios. The calculation is given by the equation

$$w_i = \frac{1}{n} \sum_{k=i}^{n} \frac{1}{k} \tag{1}$$

where n is the number of criteria, i is the ith element in the ranking, and k is used for calculate the weight of the ith element in the ranking. An Excel spreadsheet was created with the calculations required by ROC method.

2.3 Using TOPSIS

The original version of TOPSIS [1] was added to the Excel spreadsheet. The first step of TOPSIS (creation of decision matrix) was replaced by the decision matrix presented in Fig. 1. In the second step of TOPSIS it was used the vector standardization method for standardizing the decision matrix $(x_{ij})_{m \times n}$, according to equation

$$v_{ij} = \frac{x_{ij}}{\sqrt{\sum_{k=1}^{m} x_{kj}^2}} \tag{2}$$

where i = 1, 2, ..., m are the alternatives, and j = 1, 2, ..., n are the criteria that were used to evaluate the alternatives. For weighting the standardized decision matrix, the weighted arithmetic mean was used in the third step of TOPSIS for aggregating the five weighting vectors of the d = 1, ..., 5 participants in each simulation. All other steps were calculated accordingly.

2.4   Using TOPSIS for Group

The extended version of TOPSIS for group decision making [2] was also added to the Excel spreadsheet. It was used a shared decision matrix and, hence, each decision maker used the same decision matrix presented in Fig. 1. It was also used the vector standardization method for standardizing the decision matrix $(x_{ij})_{m \times n}$ and the weighted geometric mean was used in the fifth step of the extended TOPSIS for aggregating the distances between each alternative to the best and worst alternative calculated to each participant in each simulation according to equations

$$\overline{S_i^+} = \left(\prod_{d=1}^{D} S_i^{d+} \right)^{\frac{1}{b}} \tag{3}$$

and

$$\overline{S_i^-} = \left(\prod_{d=1}^{D} S_i^{d-} \right)^{\frac{1}{b}} \tag{4}$$

where S_i^{d+} and S_i^{d-} are the distances calculated for each decision maker d = 1, ..., 5, with D = 5 decision makers, between his/her *ith* alternative and the best and worst alternative, respectively. All other steps were calculated accordingly.

2.5   Using the Utility Function

The steps for the application of the utility function [3] were also added to the Excel spreadsheet, which includes the calculation of the function $\pi_d: \Re_+^D \to [0, 1]$ that represents the payoff for a player d from the set of D-players defined as

$$\pi_d \left(x_d, x_{d \neq p} \right) = \varphi(x_d, IA_d) . \prod_{d \neq p, p=1}^{D} \varphi \left(x_d, x_p \right) . \varphi \left(x_p, IA_d \right) \tag{6}$$

where $\pi_d \left(x_d, x_{d \neq p} \right)$ is the utility for the *dth* decision maker when considering swapping the alternative x_d by the subset $x_{d \neq p}$ from the set of alternatives formed by the alternatives proposed by the D − 1 remaining decision makers, and φ is given by the pairwise comparison function $\varphi: \Re_+^n \to [0, 1]$, according to equation

$$\varphi \left(x_d, x_p \right) = \left[\frac{\alpha_{x_d x_p}}{\|x_p\|} \right]^{\delta} cos\theta_{x_d x_p}, \ and \ \delta = \begin{cases} 1, if \ \alpha_{x_d x_p} \leq \|x_p\| \\ -1, \ otherwise \end{cases} \tag{7}$$

where $\alpha_{x_d x_p} = \|x_d\| \cos\theta_{x_d x_p}$ is the scalar projection of the vector x_d onto the vector x_p, and $\|x_p\| = \sqrt{\left(x_p^1\right)^2 + \left(x_p^2\right)^2 + \cdots + \left(x_p^n\right)^2}$ is the norm of the respective vector with n criteria. Intuitively, the utility function π_d is defined therefore as the potential of the dth decision maker to swap x_d to x_p, including their relativity with the ideal alternative IA_d (the utopist alternative that is composed by the best scores of each $j = 1, 2, \ldots, n$ evaluation criteria).

Finally, it was used an exhaustive search algorithm for finding pure Nash equilibrium [6] in the fifth step of the utility function procedures and, in the presence of more than one Nash equilibrium, the utilitarian principle of Harsanyi [7], which seeks to maximize the sum of the individuals' utilities, was used as a social welfare ordering in its sixth step for the selection of the equilibrium with the highest sum as the solution of the method.

2.6   Performance Evaluation

While the group was deciding the solution to the problem, the authors of the present research calculated the three MCDM methods in the Excel spreadsheet. For evaluating the performance of the different MCDM methods in predicting the outcome of group decision making, it was counting the matches between the group's agreement (if reached) with the best ranked solution of each of the MCDM method and the relative measures were compared.

3   Results and Discussion

Table 1 reports the results of the comparison of the three MCDM methods in terms of predicting the outcome of each simulated group decision making.

Table 1. Comparion of three MCDM methods

Simulation	Agreement	TOPSIS	TOPSIS for group	Utility function
1	E	E	E	E
3	D	B	B	D
5	E	E	B	E
6	E	E	E	E
8	E	E	E	E
9	E	E	E	E
10	D	E	E	E
11	E	A	E	E
12	D	E	E	A
13	D	E	E	D

(*continued*)

Table 1. (*continued*)

Simulation	Agreement	TOPSIS	TOPSIS for group	Utility function
14	E	B	B	E
15	D	E	E	E
16	D	E	E	E
17	B	E	B	E
18	E	E	E	E
19	D	E	E	D
20	D	E	E	E
21	C	E	E	D
22	E	B	B	B
23	E	E	E	E
24	C	E	E	D
25	D	C	E	D
	Matches	7	8	13
	%	32%	36%	59%

According to Table 1, there were three simulations where the group did not reached an agreement (simulations 2, 4 and 7), while the alternative B was chosen once, the alternative C was chosen twice, the alternative D was chosen nine times, and the alternative E was chosen 10 times. The alternative A was not chosen as the solution to the game in any simulation.

It is noteworthy that the results corresponds to the initial assumption that MCDM methods that does not require the application of aggregation techniques seems to provide better efficiency in predicting group decision making outcomes. Firstly, the extended TOPSIS for group decision making correctly predicted one more case than the original TOPSIS method. The result corroborates the intuition of Shih et al. [1] that "our group preferences are aggregated within the procedure [...] the results have demonstrated our model to be both robust and efficient [...] we are confident the results for various examples would give us similar conclusions". Furthermore, the MCDM method based on game theory correctly predicted five more cases than the extended version of TOPSIS method. The result corroborates the intuition of Leoneti [3] that "MCDM may have reduced efficiency due to problems with the aggregation of preferences when the decision-making process involves more than one individual [...] the advantage of using this function for modeling group multicriteria decision making problems as games is the use of game theory approach to circumvent the limitation of the aggregation procedure that is necessary for group decision making in the traditional multicriteria decision making approach".

4 Conclusions

This paper compares MCDM methods that require and do not require the use of aggregation techniques for its application within group decision making and discuss the concern regarding the underestimation of conflicts. An experiment was adapted from previous research and three MCDM methods were empirically compared based on the performance of these methods to predict correctly the group's agreement. A criterion was established to be used as a performance indicator, which was the counting of the matches between the group's agreement (if reached) with the solution of the MCDM method.

The results indicate that there is a considerable difference between the TOPSIS with a method that utilizes the concept of game theory for bypassing the aggregation step that is usual in traditional MCDM approaches. A possible conclusion, that deserves further investigations, is that modeling group decision making conflicts by using MCDM methods that do not require the use of preference aggregation techniques provides best efficiency in predicting group decision making outcomes.

Acknowledgments. The authors thank the São Paulo Research Foundation (FAPESP) for Regular Research Grant (2016/03722-5).

References

1. Shih, H.S., Shyur, H.J., Lee, E.S.: An extension of TOPSIS for group decision making. Math. Comput. Model. **45**(7–8), 801–813 (2007)
2. Hwang, C.L., Yoon, K.: Methods for multiple attribute decision making. In: Hwang, C.L., Yoon, K. (eds.) Multiple Attribute Decision Making. LNE, vol. 186, pp. 58–191. Springer, Heidelberg (1981). https://doi.org/10.1007/978-3-642-48318-9_3
3. Leoneti, A.B.: Utility function for modeling group multicriteria decision making problems as games. Oper. Res. Perspect. **3**, 21–26 (2016)
4. Leoneti, A.B., de Sessa, F.: A deviation index proposal to evaluate group decision making based on equilibrium solutions. In: Bajwa, D., Koeszegi, S., Vetschera, R. (eds.) GDN 2016. LNBIP, vol. 274, pp. 101–112. Springer, Cham (2017). https://doi.org/10.1007/978-3-319-52624-9_8
5. Barron, F.H., Barrett, B.E.: Decision quality using ranked attribute weights. Manag. Sci. **42**, 1515–1523 (1996)
6. Nash, J.: Non-cooperative games. Ann. Math. **54**, 286–295 (1951)
7. Harsanyi, J.C.: Morality and the theory of rational behavior. Soc. Res. **44**, 623–656 (1977)

Collaborative Decision Making Processes

UX Challenges in GDSS: An Experience Report

Amir Sakka[1,2(✉)] ⓘ, Gabriela Bosetti[3] ⓘ, Julián Grigera[3] ⓘ,
Guy Camilleri[1] ⓘ, Alejandro Fernández[3] ⓘ, Pascale Zaraté[1] ⓘ,
Sandro Bimonte[2] ⓘ, and Lucile Sautot[2] ⓘ

[1] IRIT, Toulouse Université, 2 rue du Doyen Gabriel Marty,
31042 Toulouse Cedex 9, France
{amir.sakka,guy.camilleri,pascale.zarate}@irit.fr
[2] IRSTEA, UR TSCF, 9 Av. B. Pascal, 63178 Aubiere, France
{sandro.bimonte,lucile.sautot}@irstea.fr
[3] LIFIA, Facultad de Informática, UNLP. 50th St. and 120 St.,
1900 La Plata, Argentina
{gabriela.bosetti,julian.grigera,
alejandro.fernandez}@lifia.info.unlp.edu.ar

Abstract. In this paper we present a user experience report on a Group Decision Support System. The used system is a Collaborative framework called GRoUp Support (GRUS). The experience consists in three user tests conducted in three different countries. While the locations are different, all three tests were run in the same conditions: same facilitator and tested process. In order to support the end-users. we teach the system in two different ways: a presentation of the system, and a video demonstrating how to use it. The main feedback of this experience is that the teaching step for using Collaborative tools in mandatory. The experience was conducted in the context of decision-making in the agriculture domain.

Keywords: GDSS · User experience

1 Introduction

The original purpose of Group Support Systems (GSS), also called Group Decision Support Systems (GDSS) is to support a group of decision-makers engaged in a decision process. This can be possible by exploiting information technology facilities. In the early 1980s, many studies started to explore how collaboration technologies (as email, chat, teleconferencing, etc.) can be used to improve the efficiency of the group work. Most of these studies focused on collaborative group decision-making and problem-solving activities.

Researchers proposed several definitions of GSS. In their work, DeSanctis and Gallupe in [1] defined GSS as a system which combines communication, computer, and decision technologies to support problem formulation and solution in group meetings. For Sprague and Carlson, a GSS is a combination of hardware, software, people and processes, that enables collaboration between groups of individuals [2].

© Springer Nature Switzerland AG 2019
D. C. Morais et al. (Eds.): GDN 2019, LNBIP 351, pp. 67–79, 2019.
https://doi.org/10.1007/978-3-030-21711-2_6

These definitions (and many others) point out four important aspects: devices (as computers, communication network), software (decision technologies, communication software, etc.), people (as the meeting participants) and group processes (as nominal group technique, etc.).

GSSs can be used in several situations:

- Face to face: decision makers are situated in a decision room,
- Distributed and synchronous situation: decision makers can work from their own office through a Web GSS and can communicate thanks to chatting system, video systems [3].

GSS (and most groupware systems) require significant investment (time and effort). They introduce changes in two dimensions: intentional group processes; and software to support them [22]. Although research on groupware systems goes back to the 60's, we still struggle to evaluate how well they fulfill their objectives (and thus, how well they justify the required investment). Evaluating groupware systems is challenging as it requires studying them from multiple perspectives: the group characteristics and its dynamics, the organizational context where the system is used, and the effects of the technology on the group's tasks [20]. Moreover, innovative GSS (especially research prototypes) frequently fail early not because of a bad design of intentional processes but because of user experience defects in the design and implementation of the supporting software. In fact, much of what is known about user experience evaluation refers to single user applications. Only recently, research has turned its attention towards user experience evaluation of groupware [21].

In this paper we report an experience using an existing GSS called GRoUp Support (GRUS). Our proposal takes profit of the user experience design of GRUS in order to improve it.

The paper is structured as follows. In the second section we describe the related work. In the third section the GRUS system is briefly explained. The fourth section describes the used protocol that is composed by a User/Interface usability test, a video design and finally the study is described. In the fifth section the three conducted user tests are described: one in Spain, another one in Argentina and the final one in Chile. The sixth section presents an analysis of the results and gives some conclusions and perspectives of this work.

2 Related Work

The original purpose of Group Support Systems (GSS), also called Group Decision Support Systems (GDSS) is to exploit opportunities that information technology tools can offer to support group work. Several definitions have been proposed which point out four important aspects: devices (computers, communication network, etc.), software (decision technologies, communication software, etc.), people (meeting participants) and group processes (as nominal group technique, etc.).

Many studies evaluate GSS and show that they can improve the productivity by increasing information flow between participants, by generating a more objective evaluation of information, by improving synergy inside the group, by reducing time, and so on (see [14, 15, 17, 19]). All these studies highlighted that the efficiency of use of GSS depends strongly on the facilitator. Group facilitation is defined as a process in which a person who is acceptable to all members of the group intervenes to help improving the way it identifies and solves problems and makes decisions [13]. Unfortunately, professional facilitators are difficult to maintain in organizations and their disappearance will often entail the abandon of use of GSS (see [16]).

In this work, we chose the Collaborative Engineering (CE) framework, which proposes for high-value recurring tasks to transfer facilitation skills to participants (in our case, to some participants that are novice in facilitation) and, in this way, to avoid to maintain professional facilitators (see [16]). As any other software application, the user experience is an important aspect for its adoption. Improving the user experience could constitute an interesting leverage for promoting the use of GSS without professional facilitators by, for example, making them more autonomous with the tool.

Usability is a crucial aspect of UX, and there are many evaluation methods to assess it. Fernandez et al. report on different usability evaluation methods (UEMs) for the web [18], many of which are applied after the system was deployed with the purpose of finding and fixing usability problems on existing web interfaces. Collaborative software presents a particular challenge for usability, either in the design process or in the user tests with many participants, which present a very different scenario than the traditional, single-user tests.

In the next section we describe the GRUS system, which was used as GSS throughout the experience.

3   The GRUS System

The system GRUS (GRoUp Support) is a Group Support System (GSS) in the form of a web application. GRUS can be used for organizing collaborative meetings in synchronous and asynchronous modes. In synchronous mode, all participants are connected to the system at the same time, while in asynchronous mode, participants use the system at different ti mes. It is also possible to use GRUS in mixed mode, synchronously and asynchronously at different steps of the process. With GRUS, users can achieve session in distributed (all participants are not in the same room) and non-distributed situations (all participants are in the same room). The only requirement is an internet connection with a web browser.

A user of GRUS can participate in several parallel meetings. For some meetings, a user can have the role of a standard participant and for other meetings she/he can have the role of a facilitator. In GRUS, the facilitator of a collaborative process can always participate to all activities of her/his process.

The GRUS system proposes several collaborative tools, the main tools are:

- *Electronic brainstorming*: allows participants to submit contributions (ideas) to the group.
- *Clustering*: the facilitator defines a set of clusters and put items inside of these clusters.
- *Vote*: This class of tools refers to voting procedure.

- *Multicriteria evaluation*: users can evaluate alternatives according to criteria.
- *Consensus*: displays statistics on the multicriteria evaluation outcomes.
- *Miscellaneous:* reporting (automatic report generation), feedback (participant questionnaire for evaluating the meeting quality), conclusion (for integrating conclusions of the meeting), direct vote (the facilitator directly assigns a value to items).

A GRUS session follows two general stages: the meeting creation and the achievement stages. During the meeting creation stage, a GRUS user defines the topic, which user will be the facilitator, the group process, the beginning date and the duration (see Fig. 1). She/he also invites participants to the meeting. A predefined process can be used or a new one can be created (see Fig. 1). In GRUS, a group process is a sequence of collaborative tools (mentioned above). The meeting is carried out in the second stage. In this stage, the facilitator starts the meeting and then, participants (including the facilitator) can contribute. The facilitator has a toolbar to manage the meeting (this toolbar is only present in the facilitator interface, see Fig. 2). Thanks to this toolbar, she/he can add/remove participants, go to the next collaborative tool (and thus finish the current step/tool), modify the group process and finish the meeting. Standard participants do not have this toolbar and only follow the sequence of steps/tools.

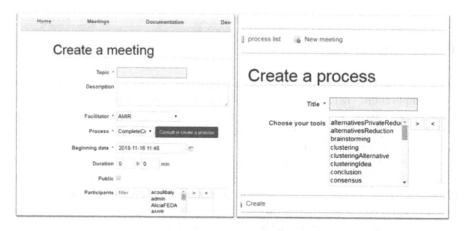

Fig. 1. Meeting and process creation in GRUS

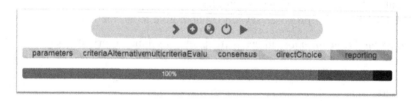

Fig. 2. GRUS: Achievement – Facilitator interface

4   Protocol

Our protocol is based on a double approach: training the users and then testing the system. Following the final step (testing) a questionnaire is given to end-users to evaluate the system as well.

The idea of this scenario allows us to re-implement it for each following training session to guarantee having the same circumstances in which the reported results would be generated. Thus, with this way the resulted feedbacks and evaluations (comments, questionnaires, etc.) of the system would be comparable because the same surveying elements might be evaluated against each other to produce a more solid and effective reporting strategy.

4.1   User Interface Evaluation

In a previous work [7] we ran a three-fold usability evaluation on GRUS, as a representative software system to support collaborative decision making. The evaluation included user tests with volunteers, and heuristic evaluation (i.e. manual inspection without users), and an automated diagnose supported by a usability service named Kobold [4].

The main aim of the evaluation was to find out the usability issues, not only of the GRUS system in particular, but also to any other Group DSS products in a more general way. The motivation behind this evaluation was to understand why, in spite of the existence of many different DSSs available for the agricultural field, the adoption rate is so marginally low. According to the literature, usability is one of the main factors for this lack of adoption [5, 6].

Being a particular setting for evaluating usability, especially given the collaborative component that's typical in such systems, the evaluation was designed with three different techniques. The motivation behind this decision was to maximize the coverage of usability issues to capture. For instance, the automated diagnose was expected to catch issues overlooked by experts, and heuristic inspection could give the experts the chance to detect issues that user tests could not cover (since tasks are designed for end-users to follow a somewhat fixed path).

Preparation. We ran the tests in the context of decision-making in the specific scenario of tomato production in the green belt of La Plata city, Argentina. For the user tests, i.e. those involving real volunteers, we designed tasks for the team to accomplish, mainly related to the different alternatives that producers face at the time of planting or harvesting. In the different tests we ran, some users were sharing the same physical space, and others were connected by video calls. More details on the preparation can be found in the previous research [7]. The automated tests were run simultaneously with the user tests, since the automated service used (Kobold) requires capturing real users' interaction in order to produce a list of usability issues.

Results. After the experiment, we detected a total of 15 issues, with some overlapping between the three different kinds of evaluation. The issues detected in the experiment were consistent with the previous findings in the literature. The most serious and repeated issues were connected with two general problems:

- Excess of information, or bloated GUIs: one representative example was the *"overloaded report"* for the decision-making process (this was actually one of the two issues captured by all three techniques). Other issues related to this general problem were *"complex GUI in multi-criteria feature selection"*, and also *"redundant controls and terminology"*.
- Lack of awareness in the collaborative process: many issues were related to the collaborative nature of the software combined with the linear process. Volunteers were frequently confused about what the next action was, or where the other participants were standing.

Many of the 15 issues were also related to the system's learnability, that is, after running into the problematic UIs for the first time, volunteers quickly learned about how to handle it. Also, the multi-step, multi-user process (assisted by a facilitator) required prior training before the usage. Because of this, we designed a video training session for further usability tests, so users could be prepared for the tasks. It was also a good way of controlling the potential bias of differently trained users.

In the next section, we present how we devised the video training so future volunteers could be quickly put up to speed in a uniform way, before running the test sessions.

4.2   Training the End Users

According to Sutcliffe and Ryan [8], one of the four techniques of the SCRAM method for requirements elicitation and validation is providing a designed artifact which users can react to, like using prototypes or conducting concept demonstrator sessions. What is presented in a demonstrator session is called a demonstrator script and its nature can vary: Røkke et al. [10] mention that it could be a prototype-simulation or even a prospective design, and the session can be interactive (with the participants using the system) or simply a presentation showing how to use the system. In any case, what is important is to generate a debate to get feedback from the participants and observe their reactions. Sutcliffe and Ryan [8] state that the demonstrator has limited functionality and interactivity, and it is intended to illustrate a typical user task. It runs "as a script" that illustrates "a scenario of typical user actions with effects mimicked by the designer". Maguire et al. [12] also mention video-prototyping as an alternative to demonstration and to show the concepts behind the system.

In our case, as we run the experience in different sessions, explained by different presenters, with participants (the audience, future users) living in different countries (Argentina, Chile, France, and Spain) and involves a collaborative scenario, we choose to present the use of the tool in a same scenario by recording a video in English and Spanish. This way, we avoided having a bias due to different environments (OS, browser, Internet connection, etc.), team configurations and order (most of the steps involve actions from different users), or differences in the explanations provided by the presenter (due to his user-experience, or the amount of details provided, possible mistakes while demonstrating each step, etc.). In addition, we generated the concept demonstrator script with two goals: to get feedback from the users but also to familiarize them with the interactions offered by the system before they had actually to use it in a different domain. In the subsections below, we detail how we generated the demonstrator and how we designed the demonstration session.

The demonstrator script was designed to show the use of GRUS [7] to make a multicriteria decision in the field of Agriculture with participants who are experts in the domain of the problem. As GRUS allows users to use predefined process and even create new ones, we choose to show its usage with the same kind of process that will be later used (after the demonstration) by the participants to solve another problem in a different domain. The process we choose was a multicriteria evaluation process, which involves: (1) defining the meeting parameters, (2) defining the criteria and alternatives, (3) multicriteria evaluation, (4) direct choice and (5) reporting.

The production of the demonstrator script started with an introduction where the scenario was explained. It was set up in the context of five greenhouse leaders from a same farm needing to agree on how much stems per plant they should use for the next crop. The farmers have this doubt since they heard from a study stating that increasing the number of stems to 3 or 4 significantly increase yield without significantly compromising the fruit quality [11]. But such experiment was run in a place where the conditions of the soil and the weather change, and with a different kind of mini-tomato seed. Such differences were presented in the introduction, as well as the five participants identified with an avatar for further references in the screen-recording sessions.

After presenting the scenario, the video presents how the five users solved the problem using GRUS. The video was divided into sections, clearly separated with a progress graph indicating which is the following step to demonstrate. The first section involved the participation of just one user who acted as a moderator and created and managed the meeting, but multiple users participated in most of the remaining steps. In such cases, the actions of the different users are presented sequentially, and their avatar was placed in a corner to indicate who is currently taking action.

To produce the demonstration part of the video, we recorded ourselves using the system and playing the role of the 5 farmers. We recorded the screens of all the users, each in his own environment. For the session recordings we defined recording guidelines, so all the participants recorded the video under the same settings: recording in mp4 format with a high resolution (720p onwards), full-screen mode, 30 fps, disabling the audio input and enabling the recording of the pointer. One of the sessions run on Ubuntu using the SimpleScreenRecorder tool[1], three on a Windows platform with the Debut video capture application[2], and one in MacOS using the QuickTime-integrated functionality[3].

The software used to produce the video was Kdenlive[4], an open-source multi-track video editor. We also defined guidelines to process the individual videos:

1. Split the steps. Identify each step of the process and render it in a separated file.
2. Split delays and remove them. If there are delays (e.g., the user is thinking), split the video to separate such sections and remove them (so the videos look fluent).

---

[1] http://www.maartenbaert.be/simplescreenrecorder/.

[2] https://www.nchsoftware.com/capture/.

[3] https://support.apple.com/guide/quicktime-player/record-your-screen-qtp97b08e666/mac.

[4] https://kdenlive.org/en/.

3. Accelerate slow actions. If there are slow sections in the video that cannot be split and removed, apply the speed filter to such Section

4. Zoom in. Apply the zoom effect when the host resolution was too high and form controls look so small, compared to the other sessions. E.g. If there is some text that the user may need to read when playing the video.

After the individual processing of the sessions, we integrated the introduction and all the individual sessions ordered by the steps in the process. We also placed a picture with the avatar of the active user in the right-bottom corner of the video. The resultant video is publicly available on Youtube.[5]

4.3   Protocol Definition

The main goal of the previously conducted experiments with GRUS using an experimentation protocol was to show how a GDSS web-based system can be supportive to a group involved in collective decision-making while being non-used to technology-based solutions to take/support critical decisions. Each of the participating teams has its own objectives, specificity of the risk to manage, and level the uncertainty to reveal.

Another important aspect of having such a unified protocol is consolidating the evaluation of user experience and satisfaction about the system's outcomes in a way that helps to get more trustworthy conclusions about what and how to enhance in the actual deployed system's features and functionalities.

To set up the user tests, the following elements must be available at the meeting room (synchronous and collocated experiment):

- Two or more decision-makers with a computer for each of them.
- Internet connection is mandatory since the system is a web-based application.
- A facilitator managing the meeting (preferably the same person in all cases or at least having equal level of system's functionalities use and explanation proficiency)
- A shared screen or a video projector to share the facilitator's screen when demonstrations are needed.

Before starting the tests, every participant must have a user account on GRUS, if not, a new one needs to be created accessing the new account form page (http://141. 115.26.26:8080/grus1112/user/newAccount/form) or by clicking on the Login button at the top right of the screen then choosing create new account.

After being logged in, the facilitator will create the new meeting with the confirmation of everyone on the parameters of the process and invite all the decision makers to join the meeting accessible from the list of meeting available in: http://141.115.26. 26:8080/grus1112/default/openMeeting or by navigating to the meetings button in the top menu.

After joining the meeting, participants must follow the facilitator instructions to fill in a proper manner all the following steps, which differs accordingly to the chosen or newly defined process.

---

[5] https://youtu.be/jkn7XhNK8hU.

When the decision to be taken is dependent on multiple influencing factors that must be considered to have a precise evaluation of the different available possibilities, then the Multi-Criteria process is the one to be followed to accomplish such a meeting. To do so, GRUS proposes a predefined MCDM process that consists of:

- Parametrization of the meeting, i.e. title, description, stakeholders' weights, evaluation scale.
- Brainstorming engaging all participants to define collectively the set of criteria and alternatives.
- Individual preferences matrix of Criteria/alternatives to be evaluated against one another done by everyone separately using the defined evaluation scale.
- Consensus building step is the one during which the facilitator shows and explains the resulting calculation results and leads the interpretation process to build up a final common decision.
- Decision to be made after having the consensus about what, based on the supportive results given by the system, is the most likely to be held as a better alternative, what might be suitable or feasible in the impossibility of the first chosen one or set of elements and what are the eliminated alternatives that had a non-encouraging score during the MC individual preferences step.
- At the end of the meeting, an automatically generated report would be downloadable containing all the parameters and results that have been used and produced during the test.

A final questionnaire with the following questions should be given to participants to get it filled after the meeting:
Evaluation of the system:

- Do you feel that the system helped making the decision?
- Do you think that the system is too complicated? In which step?
- Is the user interface user friendly?
- What would you propose as improvement of the system?

Evaluation of the previous training:

- Did the training help you understand more the system?
- Do you think that the training helped you to better define the problem?
- Do you think that the facilitator role can be enough to build the necessary understanding of the system without any prior training?
- Would one more round be enough for you to get more effectively used to the system?
- Would one more round on the same example give results that are more precise after having a better understanding of the system?

5   Results

In this section we detail the four tests that we ran: two in Spain, one in Argentina and one in Chile. As the general purpose of these sessions is evaluating the user-experience aspect of our GSS, we here only report its related users' feedback.

5.1   Tests in Spain

Two user tests were conducted in Spain in a company which is a cooperative company for the food industry.

For the first test 9 participants were involved: 2 as facilitators and 7 from as decision makers grouped on three computers, that means that we really have 3 decision makers and 1 facilitator, the second facilitator supported the end-users to use the system. We present the video in Spanish explaining how to use the GRUS system. We define together a decision problem: How to improve eggs production? 3 alternatives were defined: free range growth, growth inside a pavilion, growth kept in a cage inside a pavilion; and 3 criteria: animal welfare, type of feed, infrastructure. We used the multi-criteria process (presented below).

For the second test we had 4 computers: two computers for two end-users (decision makers), one for the decision maker and one for the facilitator. The end-users participated in the previous test. They knew the system and they learned how to use it in the first round. The video-demonstrator was not shown to them at that point. The decision to make was: *Choose a packaging for meat exportation for international countries outside Europe*. Three alternatives were defined: *Family Pack/Individual Pack/ Professional Pack: Big containers*. The alternatives were defined by discussion and the facilitator filled in the three alternatives. They finally have to order the three alternatives and the system computed the final result for the group. The used process is simpler than the previous and includes fewer steps.

Feedback from the first session showed us that the system is too complex to be used by real practitioners and that some part of the system must be hidden depending on the current step of the process. Nevertheless, they found the video very useful to understand the system and the multicriteria process as well.

The participants found the voting process very easier and faster than the multicriteria process, the presentation of results at the end as well have been also clearer and more understandable to all of them.

5.2   Test in Argentina

Our second round (third test) was held in the Faculty of Agrarian and Forestry Sciences in the National University of La Plata in Argentina with the participation of 5 decision makers, two researchers in agronomy, one researcher in computer science and two master students.

The tools to be used were defined collectively by the attendees Fig. 3a and the decision was about the allocation of the amount of each under-study crop inside the university's nursery. Therefore, because of their advanced level of knowledge and valuable experience, the agronomy researchers took very higher weights in the parametrization step. In the brainstorming step, as shown in Fig. 3b participants defined easily the set of criteria and alternatives since they understood well the operability of the system during the training session few days before. Then, in the third step the MC-evaluation gave the ranking which was the same with the two used calculation methods i.e. Choquet's integral [23] and weighted sum Fig. 3c Finally, and after discussion, decision makers reached in a consensual manner the final choice that consisted of keeping the second alternative, to consider as feasible the third and to not keep the first one as illustrated in Fig. 3d.

After the meeting the questionnaire was filled by the attendees and revealed the following:

- Users are satisfied with the features and are appreciating the assets offered by the system that simplified to them a task that was usually complicated in most similar situations.
- They consensually believe that the training session gave them an essential introduction to the system that without which, it could have been more complicated to them to define correctly the problem and to give precisely their consistent preferences.
- Some improvements from a user experience point of view were proposed, such as the revision of the matrix of preferences presentation, which might be hard to understand by non-IT user stakeholders.

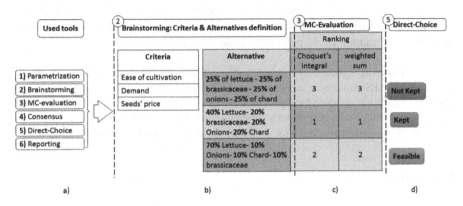

Fig. 3. Grus test in Argentina

5.3 Test in Chile

With a group of researchers and administrative staff of the main agricultural research institution in Chile, i.e. INIA La Cruz, and after a training session with the video-demonstrator, the participants collectively decided the most important topics to be prioritized by the institution during the next year.

The meeting was reached by 9 decision makers with different administrative positions and scientific backgrounds that defined together the topic and the tools to use. After discussing the level of knowledge and the influence of each on the final decision, different weights were given accordingly to everyone during the parametrization step.

Afterwards, participants defined a set of criteria (i.e. Social impact, Regional economical contribution, Regional center skills and Climate change) and alternatives i.e. Crop physiology, Irrigation research, Pollination research, Computer science application in agriculture, Agricultural ecology, Tree fruit research and Horticultural research). Next, they gave separately their personal preferences that have been collected and ranked based on Choquet's Integral and weighted sum and finally took the decision to consider both rankings by keeping the first two alternatives of each calculation method, to consider the 3rd elements as feasible and to not keep the rest.

After the meeting, the participants answered the questionnaire listed in Sect. 4.3. The level of satisfaction about the usefulness and the added-value of the system was high and consensual. The training session was helpful for most of them, nevertheless, some thought that whether only facilitation or training session would suffice, and had some difficulties with the user interface that was not sufficiently highlighting the guideline through the process.

6   Conclusions

In this paper we presented a report of user experience in a GSS. Conducting this experience in three different countries with different users (but using the same conditions in all sessions otherwise), the feedback showed us that learning such Collaborative systems is mandatory for end-users. We used for this purpose a video that simplifies the system understanding. Nevertheless, we still have to investigate our studies to understand how the training could influence the decision results.

Acknowledgements. Authors of this publication acknowledge the contribution of the Project 691249, RUC-APS: Enhancing and implementing Knowledge based ICT solutions within high Risk and Uncertain Conditions for Agriculture Production Systems (www.ruc-aps.eu), funded by the European Union under their funding scheme H2020-MSCA-RISE-2015.

This work is partially supported by the project ANR-17-CE04-0012 VGI4bio.

References

1. Desanctis, G., Gallupe, R.B.: A foundation for the study of group decision support systems. Manage. Sci. **33**(5), 589–609 (1987)
2. Sprague, R.H., Carlson, E.: Building Effective Decision Support Systems. Prentice-Hall, Englewood Cliffs (1982)
3. Camilleri, G., Zaraté, P.: How to avoid conflict in group decision making: a multicriteria approach. In: Handbook GDN. Springer (2018, in preparation)
4. Grigera, J., Garrido, A., Rossi, G.: Kobold: web usability as a service. In: Proceedings of the 32nd IEEE/ACM International Conference on Automated Software Engineering. IEEE Press (2017)
5. Rose, D.C., et al.: Decision support tools for agriculture: towards effective design and delivery. Agric. Syst. **149**, 165–174 (2016). https://doi.org/10.1016/j.agsy.2016.09.009
6. Rossi, V., Caffi, T., Salinari, F.: Helping farmers face the increasing complexity of decision-making for crop protection. Phytopathologia Mediterranea **51**(3), 457–479 (2012)
7. Grigera, J., Garrido, A., Zaraté, P., Camilleri, G., Fernández, A.: A mixed usability evaluation on a multi criteria group decision support system in agriculture. In: Proceedings of the XIX International Conference on Human Computer Interaction, p. 36. ACM, New York (2018)
8. Sutcliffe, A.G., Ryan, M.: Experience with SCRAM, a scenario requirements analysis method. In: Proceedings of IEEE Third International Conference on Requirements Engineering, pp. 164–171 (1998)

9. Sutcliffe, A.: Scenario-based requirements engineering. In: 11th IEEE International Requirements Engineering Conference, pp. 320–329 (2003)
10. Røkke, J.M., Dresser-Rand, A.S., Muller, G., Pennotti, M.: Requirement Elicitation and Validation by Prototyping and Demonstrators (2010)
11. Candian, J.S., Martins, B.N.M., Cardoso, A.I.I., Evangelista, R.M., Fujita, E.: Stem conduction systems effect on the production and quality of mini tomato under organic management. Bragantia 76(2), 238–245 (2017)
12. Maguire, M., Kirakowski, J., Vereker, N.: RESPECT: User Centred Requirements Handbook (1998)
13. Schwarz, R.: The Skilled Facilitator. Jossey-Bass Publishers, San Francisco (1994)
14. de Vreede, G.J.: Achieving repeatable team performance through collaboration engineering: experiences in two case studies. Manag. Inf. Syst. Q. Exe. 13(2), 115–129 (2014)
15. Nunamaker, J., Briggs, R., Mittleman, D., Vogel, D., Balthazard, P.: Lessons from a dozen years of group support systems research: a discussion of lab and field findings. J. Manag. Inf. Syst. 13(3), 163–207 (1996)
16. Briggs, R.O., de Vreede, G.J., Nunamaker Jr., J.F.: Collaboration engineering with thinklets to pursue sustained success with group support systems. J. Manag. Inf. Syst. 19(4), 31–64 (2003)
17. de Vreede, G.-J., Briggs, R.O.: Collaboration engineering: reflections on 15 years of research & practice. In: HICSS (2018)
18. Fernandez, A., Insfran, E., Abrahão, S.: Usability evaluation methods for the web: a systematic mapping study. Inf. Softw. Technol. 53(8), 789–817 (2011)
19. Fjermestad, J., Hiltz, S.R.: An assessment of group support systems experimental research: methodology and results. J. Manag. IS 15(3), 7–149 (1998)
20. Antunes, P., Herskovic, V., Ochoa, S.F., Pino, J.A.: Structuring dimensions for collaborative systems evaluation. ACM Comput. Surv. (2012). https://doi.org/10.1145/2089125.2089128
21. Berkman, M.İ., Karahoca, D., Karahoca, A.: A measurement and structural model for usability evaluation of shared workspace groupware. Int. J. Hum.-Comput. Interact. 34(1), 35–56 (2018). https://doi.org/10.1080/10447318.2017.1326578
22. Johnson-Lenz, P., Johnson-Lenz, T.: Groupware: the process and impacts of design choices. Computer-Mediated Communication Systems: Status and Evaluation (1982). https://doi.org/10.1093/acrefore/9780190228637.013.81
23. Grabisch, M., Murofushi, T., Sugeno, M. (eds.): Fuzzy Measures and Integrals-Theory and Applications, pp. 348–374. Physica Verlag, Heidelberg (2000)

A Voting Procedures Recommender System for Decision-Making

Adama Coulibaly[1,2]([⊠]) ⓘ, Pascale Zarate[2] ⓘ, Guy Camilleri[2] ⓘ,
Jacqueline Konate[1] ⓘ, and Fana Tangara[1] ⓘ

[1] University of Bamako/USTTB-FST, Bamako, Mali
am.coulibaly@usttb.edu.ml
[2] University of Toulouse/IRIT, Toulouse, France

Abstract. Facilitation is a critical element in decision-making using the tools of new technology. Voting is a tool commonly used in decision making. The choice of a voting procedure is not easy for a novice facilitator. So it is interesting to propose a recommendation system that assists novice facilitators in their voting procedures choice.

There are several voting procedures, some of which are difficult to explain and which can elect different options or alternatives. The best choice is one whose election is easily accepted by the group.

Voting in social choice theory is a widely studied discipline whose principles are often complex and difficult to explain at a decision-making meeting. So, a recommendation system can alleviate the facilitator on his work in finding adequate voting procedure to be applied in a group decision.

Keywords: Recommendation system · Recommender · Voting procedures · Decision-making · Facilitation tools · GRECO

1 Introduction and Background

Collective decision-making often generates conflict situations due to differences in views and interests of decision-makers about the same set of objects, hence the need for decision-support systems. Making a decision is choosing from a set of alternatives that can solve a problem in a given context (Adla 2010).

Group Decision Support Systems (GDSS) are developed to help decision makers and are most often based on computer platforms that provide decision-makers with a formal framework for reflection, and investigative skills to express the preferences and parameters of each, to evaluate them, and to provide the relevant elements for the decision-making.

This type of system consists in offering tools for group decision (Kolfschoten et al. 2007). A particular actor stands out in the process of group decision making. This is the facilitator. This actor's role is to support the group decision making. This assistance can be defined not only on the technical level, but also on the content or the decision-making process (Briggs et al. 2010). Among the tasks provided by a facilitator are:

© Springer Nature Switzerland AG 2019
D. C. Morais et al. (Eds.): GDN 2019, LNBIP 351, pp. 80–91, 2019.
https://doi.org/10.1007/978-3-030-21711-2_7

- preparation of the agenda;
- technology integration;
- technical support;
- seeking information;
- coordination of the decision-making meeting;
- recording comments and voting results;
- timing the session duration.

A usual step in these group decision-making processes is to allow each member of the group to vote. There are different voting procedures (Brams et al. 2012) that the facilitator can propose to the decision-makers. These voting procedures do not necessarily lead to the same results, provoking resistance in their acceptances. The difference in voting results depends on several factors such as the method of vote calculation, the voters number, the candidates, number, the votes way presenting.

Our goal with this work is to propose a voting system with recommendation mechanism able suggesting which procedures can be used depending the decision context.

In this paper, we will briefly introduce the recommendation systems and mention some facilitation tools. In addition, we are interested in certain parameters that can influence a voting procedures results. Then our article approaches the voting theory in order to understand the procedures and the different paradoxes that can arise. We will try to understand the design of a recommendation system. As scientific contribution, we will propose a voting recommender system for a facilitator to help him in his task.

2 Related Work

2.1 Recommendation System

Recommendation systems (RS) are software tools and techniques that provide suggestions for articles that are useful to the user (Ricci et al. 2015) Suggestions focus on various decision-making processes, such as which articles to buy, which music to listen to, which online news to read, which method to choose, etc. They therefore have the potential to support and improve the quality of decisions made by users. There are four main families of recommendation systems:

Collaborative filtering is a method of making automatic predictions about the interests of a user by collecting preferences or taste information from many users. The assumption of the collaborative filtering approach is that if a person A has the same opinion as a person B on an issue, A is more likely to have B's opinion on a different issue than that of a randomly chosen person. The techniques of this approach are grouped into two subgroups: Memory-based, Model-based (Felfernig et al. 2006).

Content-based recommendation systems analyze item descriptions to identify items that are of particular interest to the user. This kind of system is composed of three main components: A Content Analyzer, that give us a classification of the items, using some sort of representation, A Profile Learner, that makes a profile that represents each user's preferences and A Filtering Component, that takes all the inputs and generates the list of recommendations for each user. But this method also has disadvantages. To make recommendations in relation to user preferences, the user must be familiar with the system. Thus during the initialization step of the preferences of the user, the system will not be able to make recommendations or these will be irrelevant.

Knowledge-based recommenders are a specific type of recommender system that are based on explicit knowledge about the item assortment, user preferences, and recommendation criteria (i.e., which item should be recommended in which context). These systems are applied in scenarios where alternative approaches such as collaborative filtering and content-based filtering cannot be applied.

A major strength of knowledge-based recommender systems is the non-existence of cold-start (ramp-up) problems. A corresponding drawback is a potential knowledge acquisition bottleneck triggered by the need to define recommendation knowledge in an explicit fashion. (Burke 2000).

Hybrid recommendation is a combination of content-based, collaborative and Knowledge-based recommendations. The aim is to eliminate the disadvantages of the tree approaches. There are different hybridization designs: Monolithic (exploiting different features), Parallel (use of several systems) and Pipelined (invocation of different systems).

For more details on the recommendation systems I can consult the works (Resnick et al. 1997, Jannach et al. 2010, Felfernig et al. 2006).

2.2 GDSS Tools

Facilitation is an important and difficult task in making a decision, so the use of computer tools is advisable. Currently several solutions exist i.e. **Stormz**[1], **Mentimeter**[2], **Sli.do**[3], **SessionLab**[4], **Howspace**[5], etc.

Some offer voting tools that only use plurality as a method of calculating votes. Our approach is to offer a tool with several procedures (such as Condorcet, Borda, etc.) accompanied by a recommendation depending on the context to accompany a facilitator.

2.3 Voting Theory

A voting procedure consists of determining from a method the winner of a vote. This gives voting procedures the character of decision-making tools in a context of social choice; whose purpose is, not only to elect a winner(s) but to build objectively a collective choice (Craid 2016). There are several voting procedures that have emerged based on specific situations. In the literature we can group these procedures into two groups namely the non-ranked (Plurality Voting, plurality with Runoff Voting, Approval Voting) and ranked procedures. Ranked procedures can also be divided into two subgroups: Not Condorcet–Consistent (Borda's count, Alternative vote, Coombs'

---

[1] https://stormz.me/en/stormz-application.

[2] https://www.mentimeter.com/.

[3] https://www.sli.do/.

[4] https://www.sessionlab.com/.

[5] https://www.howspace.com/.

method, Bucklin's method, Range voting, majority Judgement) and Ranked Condorcet–Consistent (Minimax, Dodgson, Nason, Copeland, Black, Kemeny, Schwart, Yong) (Felsenthal et al. 2018). A procedure is called Condorcet-consistent (RCC) if, as soon as there is a Condorcet winner[6] for a profile, the rule designates him as the sole winner of the election. And, it says Not Condorcet–Consistent (RNC), if it can designate other winners besides that of Condorcet. Thus all procedures derived from the Condorcet method are RCC.

We designate the set of voting procedures by VP_i,

$$VP_i = \{i | i \in [Copeland, Borda, Approval . . . MJ]\}$$

All these procedures have shown their limit in a given situation, called paradox in the voting theory (Nurmi 2012; Felsenthal et al. 2018). We define the "voting paradox" as an undesirable result that a voting procedure may produce and which may at first glance be seen, at least by some people, as surprising or counterintuitive. These paradoxes have been well studied for decades. The conclusions reached by its various studies have allowed to distinguish between two types of voting paradoxes associated with a given voting procedure: 'Simple or Straightforward' paradoxes and 'Conditional' paradoxes.

Relevant data that may influence the results of a vote are: the number of voters, the number of candidates that must be elected, the preference ordering of every voter among the competing candidates, the amount of information voters have regarding all other voters'preference orderings, the order in which voters cast their votes if it is not simultaneous, the order in which candidates are voted upon if candidates are not voted upon simultaneously, whether voting is open or secret, and the manner in which ties are to be broken (Nurmi 2012).

The five best–known 'simple 'paradoxes that may afflict voting procedures designed to elect one out of two or more candidates are the following: Condorcet Winner, Absolute Majority Winner, Condorcet Loser or Borda Paradox, Absolute Majority Loser, Pareto (or Dominated Candidate). For more information, see (Felsenthal et al. 2018; Cheng et al. 2012).

As conditional paradoxes that can influence the results of a voting procedure we can quote: Additional Support (or Lack of Monotonicity or Negative Responsiveness), Reinforcement (or Inconsistency or Multiple Districts), Truncation, no–Show, Twin, Violation of the Subset Choice Condition (SCC), Preference Inversion, Dependence on Order of Voting (DOV) see (Felsenthal et al. 2018; Nurmi 2012) for more information.

We designate the set of paradoxes by Pdx_j,

where $Pdx_j = \{j | j \in [Condorcet \ winner, . . DOV]\}$.

---

[6] http://www.whydomath.org/node/voting/impossible.html.

3   GRECO (Group vote RECOmmendation)

Our goal is to provide a hybrid recommendation engine, using voting procedures characterization for the content based approach. After also doing collaborative filtering when the information will grow. As shown in the following Fig. 1 (Inspired by work (Jannach et al. 2010)).

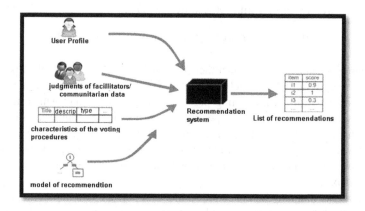

Fig. 1. GRECO recommendation logic

Currently, we have implemented voting procedures such as Borda, Condorcet, plurality, Black and Copeland methods.

3.1   Characterization of Voting Procedures

Based on the characterization of voting procedures on the following work (Suitt et al. 2014, Nurmi 2012, Durand 2000, Konczak and Lang 2005, Felsenthal et al. 2018), and taking into account that our system aims at a reduced work group environment, we have established a matrix characteristic of the implemented procedures. For a small group, we have established the following criteria: $C1$: Condorcet Winner Criterion, $C2$: Absolute Majority Criterion, $C3$: Pareto Criterion, $C4$: Loser Criterion, $C5$: Participation Criterion, $C6$: Monotony Criterion, $C7$: Coherence Criterion.

Thus, we obtain the following characterization matrix (Table 1).

Table 1. Voting procedures characterization matrix: M_P

Characteristic→ Procedures	Type	C_1	C_2	C_3	C_4	C_5	C_6	C_7
Plurality	RNC	1	2	0	2	0	0	0
Borda	RNC	1	2	0	0	0	0	0
Condorcet	RCC	1	0	0	0	1	0	1
Black	RCC	0	0	0	0	1	0	1
Copeland	RCC	0	0	0	0	1	0	1

Considering the following scoring scale, 0: the criterion does not affect the voting procedure, 1: the criterion affects the voting procedure and 2: the criterion has a significant impact on the procedure.

3.2 Voting Procedures Scoring

In GRECO, as feedback, the facilitator has the possibility to assign or evaluate the voting procedures used in a decision making process. They can use the following ratings (Table 2).

Table 2. Rating scale

Rating	Poor	Not enough	Fair	Satisfying	Good	Great
Note	0	1	2	3	4	5

We have the following scenarios for scoring the different voting procedures:

Scenarios 1: The facilitator may decide to apply a given procedure, i.e. manual selection. This choice implies that the procedure is known and appreciated by the facilitator. If the facilitator confirms his choice, the system assigns a **'Good'** rating to the chosen procedure. This mechanism allows the system to avoid the start-up problem in issuing recommendations known as 'Cold-Start'.

Scenario 2: The system can automatically propose to the facilitator a list of voting procedures to be applied. If the facilitator confirms his choice, the system assigns a 'Good' rating to the chosen procedure.

Scenario 3: After a voting procedure has been applied in a given context, participants in decision-making can address the group's overall level of satisfaction to the facilitator. This makes it possible to note the procedure used. This note is very critical and important because it comes from the group of decision-makers.

The various facilitators' notes make it possible to draw up an M_n matrix, containing voting procedures scoring as shown in Table 3.

Table 3. M_n, voting procedure scoring matrix

Procedure → Facilitator	Borda	Condorcet	Black	Pluralité	Copeland
Fac_1	5	5		3	
Fac_2			3		5
Fac_3		4			
......					
Fac_n	3				4

3.3   GRECO's Content-Based Implementation Algorithm

We used the Django framework to develop our solution. This framework is based on python and closes libraries such as Pandas[7], nump[8], scipy[9] which facilitate the implementation of the various desired functionalities. The following algorithm explains the draft of our content-based recommendation.

```
Algorithm
Data: Mp: Voting procedures characterization matrix
         Mn : Voting procedures scoring matrix
Begin
   1. build a user profile based on the voting procedures
   already used in past meetings using Mp  and   Mn
      1.1-Center the score matrix to get  Mc  ←Mn
      1.2 Calculation of  the  coordinates  for  each charac-
   teristic
   2. search for the k voting procedure profiles most sim-
   ilar to the user profile
      2.1- Index each voting procedure by its characteris-
   tics
      2.2- Look for the k profiles of the voting procedures
   most similar to the user profile using the vector model
   (Cosine similarity[10])
End
```

3.4   Using Greco: Practical Test

An example will allow us to discover the current state of GREO. For example, a committee of five (5) decision-makers wants to choose a place to celebrate the annual board of directors. Three (3) hotels (Azalaï, Grand Mic asa, Radisson Blu) have been proposed. The meeting used GRECO to determine the elected hotel according to the table containing the preferences issued by the committee.

Table 4. Voters preferences

Nb DM→	2	2	1
Rank	Radisson Blu	Grand Micasa	Radisson Blu
	Azalaï	Azalaï	Grand Micasa
	Grand Micasa	Radisson Blu	Azalaï

---

[7] https://pandas.pydata.org/.

[8] http://www.numpy.org/.

[9] https://www.scipy.org/.

The vote creation on GRECO is done in three essential steps:

Step 1: Vote creating (see Fig. 2).

a. All the basic information of the vote is provided: title, description, start and end dates of the vote and status.
b. The different candidates from the list of alternatives proposed during the meeting are added.
c. The voters who are participating in the meeting are designated and click on the button "Create the vote".

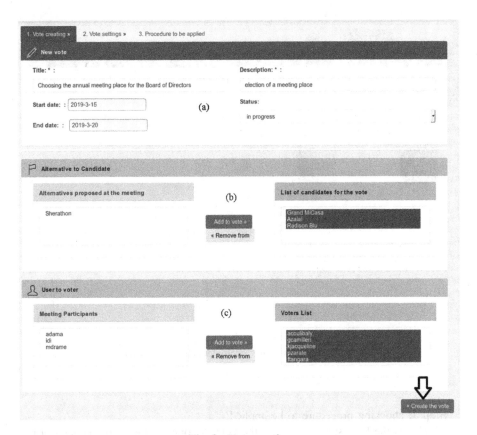

Fig. 2. Vote creating

Step 2: Vote settings (See Fig. 3).

a. We continue with a summary of the voting data during the creation process. (voting data).

b. The parameters for the recommendation are defined:

The type of procedure which is a list composed of three values (no matter, Condorcet-Consistent and Not Condorcet-Consistent). This parameter allows us to define the rank of similarity search for recommendation result. In this example, we chose 'Condorcet-Consistent'.

The parameter of choice methods takes two possible values (automatic or manual) and allows to refine the result of our recommendation because the similarity can give us a list of procedures corresponding to the user's profile. In this example we choose 'Automatic'.

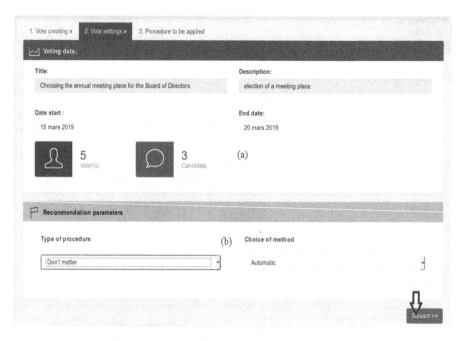

Fig. 3. Voting setting: recommendation parameters

Step 3: choosing procedure to be applied:

This is the final step in creating the vote. It confirms the recommended procedure by associating it with the vote being created. In this example the recommendation suggested the Condorcet procedure. The "Finish" button allows you to finalize the voting creation process.

Once the vote has been created, all voters can participate by making their preference list as shown in Table 4.

Finally, the Fig. 4 shows the voting result using the Condorcet procedure, and the candidate hotel '*Radisson Blu*' is the winner.

Results of the vote:
Title:
Choosing the annual meeting place for the Board of Directors
Description:
election of a meeting place
Dete start :
15 mars 2019
End date:
20 mars 2019
Condorcet's Method
Numerical
Azalaï > Grand MiCasa: 2 point(s)
Grand MiCasa > Azalaï: 3 point(s)
Azalaï > Radison Blu: 2 point(s)
Radison Blu > Azalaï: 3 point(s)
Grand MiCasa > Radison Blu: 2 point(s)
Radison Blu > Grand MiCasa: 3 point(s)
the winner is Radison Blu
Graphical
★voting procedure Scoring

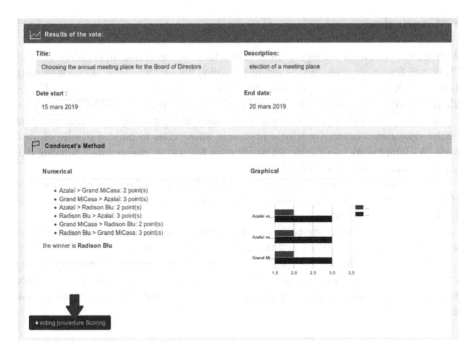

Fig. 4. Voting result

By clicking on the "***voting procedure scoring***" button, the facilitator can express the level of satisfaction of participants in the decision-making process with the voting procedure used by entering one note and a comment as shown in Fig. 5.

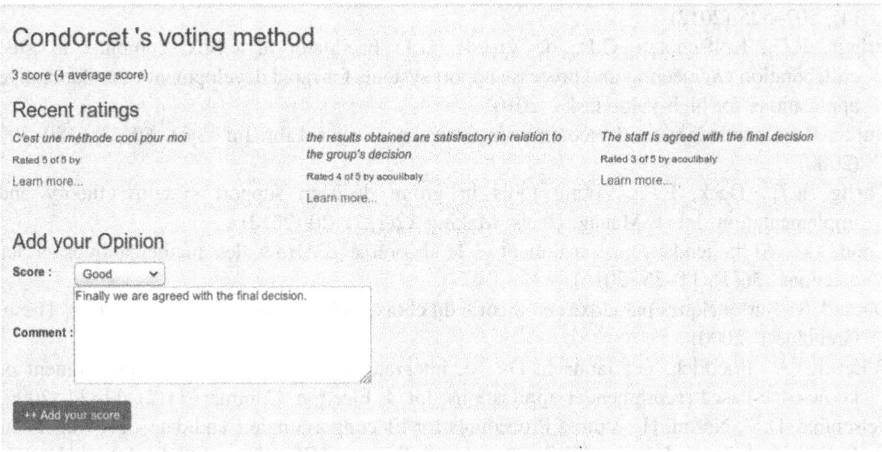

Condorcet 's voting method
3 score (4 average score)
Recent ratings
C'est une méthode cool pour moi
Rated 5 of 5 by
Learn more...
the results obtained are satisfactory in relation to the group's decision
Rated 4 of 5 by acoulibaly
Learn more...
The staff is agreed with the final decision
Rated 3 of 5 by acoulibaly
Learn more...
Add your Opinion
Score :
Good
Comment :
Finally we are agreed with the final decision.
++ Add your score

Fig. 5. Voting procedure scoring

4 Conclusions

With the use of new technologies, the role of a facilitator is crucial in decision-making. There are few tools that can recommend voting procedures in a decision-making meeting. **GRECO** comes to fill one this rarity. At current state, Condorcet, Borda, Black, plurality, Copeland voting procedures are implemented in **GRECO**.

We can conclude that the voting procedures the paradoxes and recommender system operations, especially hybrid approach, allowed us to build our solution proposal. As future work, we continued to improve the part of collaborative filtering that requires usage information in the system.

We are planning additions to other methods to have a lot of possibilities at the time of the recommendation. We recommend doing several tests to validate the results of our recommendation system.

Our recommendation engine is based essentially on the relationships that exist between the voting procedures and the mentioned paradoxes, some of which are circumstantial. In perceptive, we propose to do a study showing a ranking of the importance of their influence in the voting results. This will make it possible to reduce the number of variables in the similarity calculations thus making the recommendation faster.

References

Adla, A.: Aide à la Facilitation pour une prise de Décision Collective: Proposition d'un Modèle et d'un Outil. Thèse doctorat, Université de Toulouse, Toulous (2010)

Aggarwal, C.C.: Recommender Systems: The Textbook, pp. 1–28. Springer, Cham (2016). https://doi.org/10.1007/978-3-319-29659-3

Benouaret, I.: Un système de recommandation contextuel et composite pour la visite personnalisée de sites culturels (2017)

Brams, S.J., Kilgour, D.M.: Narrowing the field in elections: the next-two rule. J. Theor. Polit. **24**(4), 507–525 (2012)

Briggs, R.O., Kolfschoten, G.L., de Vreede, G.J.: Facilitator in a box: computer assisted collaboration engineering and process support systems for rapid development of collaborative applications for high-value tasks (2010)

Burke, R.: Knowledge-based recommender systems. Encycl. Libr. Inf. Syst. **69**(32), 180–200 (2000)

Cheng, K.E., Deek, F.P.: Voting tools in group decision support systems: theory and implementation. Int. J. Manag. Decis. Making **12**(1), 1–20 (2012)

Craid, G.: Si la tendance se maintient... le théorème d'Arrow, les mathématiques et les élections **56**(3), 11–26 (2016)

Durand, S.: Sur quelques paradoxes en théorie du choix social et en décision multicritère. Thèse. Grenoble I (2000)

Felfernig, A., Friedrich, G., Jannach, D.: An integrated environment for the development of knowledge-based recommender applications. Int. J. Electron. Commer. **11**(2), 11–34 (2006)

Felsenthal, D.S., Nurmi, H.: Voting Procedures for Electing a Single Candidate - Proving Their (In)Vulnerability to Various Voting Paradoxes. Springer, Heidelberg (2018). https://doi.org/10.1007/978-3-319-74033-1

Fomba, S.: Décision Multicritère: Un système de recommandation pour le choix de l'operateur d'agregation. Toulouse: s.n. (2018)

Fomba, S., et al.: A recommender system based on multi-criteria aggregation. Int. J. Decis. Support Syst. Technol. (IJDSST) **9**(4), 1–15 (2017)

Jannach, D., Zanker, M., Felfernig, A.: Recommender Systems: An Introduction. Cambridge University, Cambridge (2010)

Kolfschoten, G.L., Den Hengst-Bruggeling, M., De Vreede, G.-J.: Issues in the design of facilitated collaboration. Group Decis. Negot. **16**(4), 347–361 (2007)

Konczak, K., Lang, J.: Voting procedures with incomplete preferences. In: Proceedings IJCAI-05 Multidisciplinary Workshop on Advances in Preference Handling (2005)

Nurmi, H.: On the relevance of theoretical results to voting system choice. In: Felsenthal, D., Machover, M. (eds.) Electoral Systems. Studies in Choice and Welfare, pp. 255–274. Springer, Heidelberg (2012). https://doi.org/10.1007/978-3-642-20441-8_10

Resnick, P., Varian, H.R.: Recommender systems. Commun. ACM **40**(3), 56–59 (1997)

Ricci, F., Rokach, L., Shapira, B.: Recommender systems: introduction and challenges. In: Ricci, F., Rokach, L., Shapira, B. (eds.) Recommender Systems Handbook, pp. 1–34. Springer, Boston (2015). https://doi.org/10.1007/978-1-4899-7637-6_1

Suitt, J.G., Guyon, A., Hennion, T.: Vers un système de vote plus juste? Hal (2014)

Why Is It Worth It to Expand Your Set of Objectives? Impacts from Behavioral Decision Analysis in Action

Valentina Ferretti[1,2(✉)] ⓘD

[1] ABC Department, Politecnico di Milano, Via Giuseppe Ponzio 31,
20133 Milan, Italy
valentinal.ferretti@polimi.it
[2] London School of Economics and Political Science, London WC2A 2AE, UK

Abstract. The generation of objectives is a crucial component of any decision-making process. However, research has shown that Decision Makers, if unaided, are considerably deficient in utilizing personal knowledge and values to form objectives for the decisions they face. This paper discusses two real interventions in which the author employed value-focused thinking devices to support both private and public organizations in generating objectives within strategic decision-making processes. The two projects deal with the challenging task of defining a new regional transportation plan for 2050 in Italy and the design of a novel selection process for a Foundation in Hungary that provides educational opportunities for underprivileged children, respectively. The aim of this paper is to share lessons learned from prescriptive interventions and discuss the impacts of collaborative Behavioral Decision Analysis in challenging societal contexts. The contribution provided by this study feeds the debate on boosting decision making by highlighting how process boosts can improve subjects' active competences for better and more empowered strategic decisions.

Keywords: Collaborative decision making · Action research · Decision quality

1 Introduction

When we are confronted with a decision problem or opportunity, our ultimate desire is to ovoid undesirable consequences and to achieve desirable ones [1]. However, if we do not know or think about which desirable consequences we want to achieve, i.e. how our destination should look like, we may not reach it, or we may not recognize it when we get there [2].

Desires and values of decision makers are made explicit with objectives, usually expressed by a verb and an object (e.g. minimize soil pollution, maximize safety, minimize distance to the subway station, etc.). Generating a comprehensive list of relevant objectives during the framing of a decision is thus a crucial step for decision quality, as highlighted in the literature since the 18th century [e.g. 3, 4]. The main purposes for which objectives are usually used are: clarifying why one cares about the decision, stimulating the creation of better alternatives compared to the readily available ones, incorporating multiple stakeholders' views, describing the consequences of

© Springer Nature Switzerland AG 2019
D. C. Morais et al. (Eds.): GDN 2019, LNBIP 351, pp. 92–105, 2019.
https://doi.org/10.1007/978-3-030-21711-2_8

alternatives, helping screen out proposed alternatives that are non-contenders, providing a basis for evaluating the alternatives that remain, and communicating the pros and cons of the alternatives to relevant parties [5]. Defining objectives for the above mentioned purposes has been shown to enhance understanding and interest in the decision problem and increase commitment to act [5].

What we know from behavioral science is that when we have to think about objectives for a specific decision we are facing, even with significant personal consequences, we miss nearly half of the objectives that we will later acknowledge to be relevant. More worryingly, omitted objectives are usually perceived to be as important as those we are able to generate on our own. Decision Makers are thus considerably deficient in utilizing personal knowledge and values to form objectives for the decisions they face [6].

Two possible reasons may explain this impediment: not thinking broadly enough about the range of relevant objectives, and not thinking deeply enough to articulate every objective within the range that is considered [5]. Moreover, research on time-relevant decisions and "temporal construal" shows that decision makers often focus only on either the near future or distant future, to the exclusion of the other [e.g. 7]. As a consequence, individuals usually identify either short-term or long-term objectives due to an inherent inability to cross the mental boundary that separates these categories of objectives. Fortunately, research on memory also suggests mechanisms that may enable individuals to cross category boundaries during memory retrieval. For example, the provision of category names can help to recall items subordinate to those categories [e.g. 8].

This talk will discuss two real interventions in which the author used Value Focused Thinking devices [9] to support both public and private organizations in the generation of a comprehensive set of relevant objectives to be used in strategic decision-making processes. The two projects deal with the challenging task of defining a new transportation plan for 2050 for a Region in Italy, and the design of a novel selection process for a Foundation in Hungary that provides educational opportunities for underprivileged children, respectively.

The aim of the study is to share lessons learned from prescriptive interventions and discuss the impacts of Behavioral Decision Analysis in challenging societal contexts.

The contribution provided by this study feeds the debate on boosting decision making [10] by highlighting how process boosts can improve subjects' active competences for better and more empowered strategic decisions.

The reminder of the paper is organized as follows. Section 2 illustrates the methodological approach employed in the interventions and describes prior work relevant to the generation of objectives. Section 3 discusses the impacts from two behavioral decision analysis interventions by providing for each of them the contextualization of the decision environment and a discussion of the lessons learned. Finally, Sect. 4 concludes the paper by summarizing key implications and envisioning future directions of research.

2 A Process Boost: Value Focused Thinking Devices

Whether one wishes to help a decision maker with clear qualitative thinking or with a quantitative analysis, a key element in the initial divergent thinking phase of the decision process consists in the identification of a comprehensive set of objectives.

Research has shown that consulting a "master list" of objectives provides useful support for decision makers facing personal or professional decisions [e.g. 5, 6]. If such a template does exist, it should be consulted, but only after the individual has independently deliberated about his or her own objectives [5]. However, for most strategic decisions, i.e. decisions that are important in terms of consequences and the precedent they set, a "master list" of relevant objectives is unlikely to be available.

When such a template does not exist, the key to identifying decision objectives is asking the right questions to the actors involved in the decision. Keeney [9] has identified 10 categories of questions that can be asked to help identify decision objectives (Table 1). These questions should be tailored to the problem and to the individual being interviewed, the group being facilitated, or the survey being designed. For example, the strategic objectives question might be posed to the senior decision maker in an interview, while the consequences question may be posed to key stakeholders in a facilitated group [11].

The reason for the selection of Value Focused Thinking devices [9] in this paper is linked to the applicability of the technique to any decision, with or without the guidance of a professional (e.g. facilitator or consultant).

Other prescriptive approaches to stimulate the generation of a more comprehensive set of objectives have also been discussed in the literature. For example, Bond et al. [5] tested the following three approaches for the generation of objectives: (i) the provision of sample objectives, (ii) the organisation of objectives by category (i.e. means to encourage the decision makers to think more broadly), and (iii) direct challenges to do better, with or without a warning that important objectives are missing (i.e. means to encourage decision makers to think more deeply). The use of category names and direct challenges with a warning both led to improvements in the quantity as well as in the quality of objectives generated [5]. Challenging individual with a specific level of expected improvement (e.g. asking the decision maker to generate 3-6-9 more objectives) lead to significant improvement in the number of objectives, consistently with motivation theory, but individuals must first be convinced that their initial generation is incomplete and motivate to make an additional effort [5]. Specific preselected alternatives may also be used as prompts to generate objectives [1]. In other cases, decision makers may be asked to identify their fundamental objectives by reflecting on pre-specified attributes [12]. Another interesting and well consolidated approach consists in using cognitive mapping to support both the generation of objectives and a better understanding of the decision at stake [e.g. 13–15]. Finally, promising insights for the generation of objectives come also from the application of SWOT analysis [e.g. 16] and stakeholders' analysis [e.g. 17, 18].

Table 1. Techniques to use in identifying objectives [9, 11]

Category of questions	Specific questions
1. A wish list	What do you want? What do you value? What should you want? What are you trying to achieve? If money was not an obstacle, what would you do?
2. Alternatives	What is a perfect alternative, a terrible alternative, some reasonable alternative? What is good or bad about each?
3. Problems and shortcomings	What is wrong or right with your organization or enterprise? What needs fixing? What are the capability, product, or service gaps that exist?
4. Consequences	What has occurred that was good or bad? What might occur that you care about? What are the potential risks you face? What are the best or worst consequences that could occur? What could cause these?
5. Goals, constraints and guidelines	What are your aspirations? What limitations are placed upon you? Are there any legal, organizational, technological, social or political constraints?
6. Different perspectives	What would your competitor or your constituency be concerned about? At some time in the future, what would concern you? What do your stakeholders want? What do your customers want? What do your adversaries want?
7. Strategic objectives	What are your ultimate or long-range objectives? What are your values that are absolutely fundamental? What is your strategy to achieve these objectives?
8. Generic objectives	What objectives do you have for your customers, your employees, your shareholders, yourself? What environmental, social, economic, or health and safety objectives are important?
9. Structuring objectives	Follow means-ends relationships: why is that objective important, how can you achieve it? Use specification: what do you mean by this objective?
10. Quantifying objectives	How would you measure achievement of this objective? Why is objective A three times as important as objective B?

3 Impacts from Two Behavioral Decision Analysis Interventions

This section presents two projects in which the author employed the value focused thinking devices illustrated in Table 1. The two projects have been selected among the recent interventions led by the author of this paper as they allow to compare and discuss impacts of the objectives' generation devices in two different contexts, i.e. public policy making and private foundations, respectively.

For both interventions, the following paragraphs will provide the contextualization of the decision environment as well as a discussion of the results and lessons learned within the project.

3.1   The Challenging Task of Defining a New Regional Transportation Plan for 2050

Context of the Intervention. In 2014 a leading Region[1] in Northern Italy asked the SiTI research Institute[2] in Turin to support them in the preparation of the new regional transportation plan for 2050. SiTI (Higher Institute on Territorial Systems for Innovation) is a non-profit association set up in 2002 between the Politecnico di Torino and the Compagnia di San Paolo, to carry out research and training oriented towards innovation and socio-economic growth. The Institute has extensive experience in supporting public administrations and authorities at different levels, ranging from the municipal to the national one, in the following sectors: logistics and transport, environmental heritage and urban redevelopment, environmental protection, safety and social housing.

The Region under analysis, i.e. the client in this project, had two key initial questions:

(i)   How can we be sure that the set of objectives we have in mind for the new regional transportation plan for 2050 is a good one? In other words, how can we be sure that we are not missing any relevant objective?

(ii) Given that we have limited financial resources and we want these objectives to lead to a set of concrete actions for the territory, how can we understand what are the most relevant objectives from which to start? In other words, how can we generate agreement on the priorities of the plan?

A Regional Transportation Plan aims at providing the public administration with suitable tools for facing citizens' and enterprises new needs, in a logic of anticipation and not emergency response. In particular, the plan should promote innovation in the transportation system by ensuring effective governance across all dimensions of sustainable development, i.e. economic, social and environmental.

As a consequence, the definition of a new regional transportation plan represents a strategic decision making problem characterized by: long time horizons and therefore high levels of uncertainty, the contemporary presence of multiple and conflicting objectives, the need to take into account and involve multiple stakeholders and decision makers, the generation of important consequences and the inherent irreversible allocation of public resources, which calls for a transparent and accountable decision making process.

To address the client's requests, we developed a facilitated decision-making process organized according to the following key phases: framing, structuring and eliciting, and finally discussing the results. We organized a collaborative workshop for each key phase in the process and invited two representatives for each sector of the regional transportation authority, i.e. planning, info-mobility, rail networks and airports, road

---

[1] The name of the Region cannot be disclosed due to a confidentiality agreement with the local authority for this project.

[2] www.siti.polito.it.

access, road security, public transportation services, logistics and navigation, strategic infrastructures (Fig. 1).

During the first full day workshop we supported the participants in using the value focused thinking devices presented in Table 1 to generate a shared and comprehensive list of relevant objectives for the new transportation plan. The results, presented in the following section, became the input to the second workshop during which we used Multi Criteria Decision Analysis tools (MCDA) [e.g. 13] to elicit preference information from the involved experts on the relative importance of the considered objectives. Finally, during the last full day workshop, results were fed back to the involved experts for a final discussion on strategic actions to be developed for the effective achievement of each objective.

Fig. 1. Key moments during the first and third collaborative workshops

Results. To understand the impact of the first stage of this intervention, it is helpful to compare the client's initial mental model with the new set of objectives generated after using the devices presented in Table 1 and adapted to this specific decision-making environment. The client's initial view on the new regional transportation plan is summarized in Table 2.

Table 2. The client's initial view about the new transportation plan

2 guiding ideas	Quality of life and economic development
4 strategic objectives	Safety, accessibility, dynamic economic development, innovation
12 operational objectives	Improvement of alternative transportation modes; availability of information; emissions' reduction; improving the quality of the local transportation system; improving train network safety; generation of positive territorial impact from big infrastructural projects; ensuring territorial continuity for infrastructural projects; improving multi-modal links in the network for both people and goods; developing internalization policies for transportation costs; fostering the development of enterprises in the transportation sector; improving transparency in the use of resources; development of educational activities

Table 3 shows the result of the Value Focused Thinking collaborative workshop. Each participant independently developed a list of objectives for the decision, by answering on his or her own the key questions presented during the workshop. Then group discussion occurred. Similar objectives were grouped together, and duplicates omitted.

Table 3. The new set of agreed objectives for the transportation plan for 2050

To decrease the number of accidents (safety)
To improve the security of people and goods
To promote car sharing for people mobility
To promote the use of the rail and water transportation systems for the transportation of goods
To improve the offer and the characteristics of car sharing
To promote the use of low impact transportation options (electric vehicles, hybrid systems, bicycles, …)
To improve the efficacy of the transportation system satisfying the expectations of citizens
To plan and implement accessibility actions for important transportation nodes
To guarantee clear, complete and accessible information
To guarantee reliable journey times both for goods and for people
To complete and enhance functional nodes and infrastructural and technological networks
To plan an integrated transportation system for people and goods
To decrease transportation costs
To decrease public contribution to local public transportation
To promote the development of logistic activities with added value
To develop services and infrastructures for a sustainable tourism
To promote the development of innovative technologies in the transportation field
To increase positive impacts on the territories crossed by strategic infrastructures
To minimize the use of soil for new infrastructures and logistic activities
To regenerate urban areas (emission reduction, congestion charging, etc.)
To promote bicycle and pedestrian mobility within urban areas

Lessons Learned. From the comparison between Tables 2 and 3, we can notice two important effects. First, the set of objectives has significantly expanded as a result of the value focused thinking collaborative workshop (from 12 operational objectives in the client's initial mental model to 21 agreed objectives as a result of the process). Second, the quality of the objectives has also improved as they became more specific (all expressed with a verb and an object), less ambiguous and therefore also measurable.

Moreover, the results of the second workshop highlighted the strategic importance of the following top ranked objectives: (i) "improving the efficacy of the transportation system, satisfying the expectations of citizens", (ii) "to plan an integrated transportation system for people and goods", (iii) "to promote the development of logistic activities with added value" and (iv) "to promote the development of innovative technologies in the transportation field". Among these top ranked objectives, items i, iii, and iv were not among the original objectives considered by the client in their initial view of the future transportation plan. These results confirm that not only we miss objectives when facing a decision unaided, but the objectives that are missed are also very important.

The feedback reported by the participants in the collaborative decision-making process (Fig. 1) highlighted that the questions that were more effective in helping them to identify key objectives were those about alternatives, problems and shortcomings, goals, constraints and guidelines, strategic objectives, as well as the use of different perspectives throughout the process.

The key impacts observed after this behavioral decision analysis intervention can be summarized as follows:

(i)   generation of a broader and more specific set of concrete actions associated to the most important objectives compared to the set that would have been obtained based on the client's initial mental model;

(ii) consensus building among the participants in the process: having generated, discussed and agreed together a shared list of relevant objectives for the new transportation plan reduced the length of the subsequent phases in the process compared to previous projects, as agreement on the objectives to be achieved generated clarity and commitment for action;

(iii) capacity building: the organization has learned an approach that allows for a versatile and case specific use, thus constituting a working tool for the public administration that can be reused for any choice problem among competing alternatives.

3.2 The Design of a Novel Selection Process for a Foundation in Hungary that Provides Educational Opportunities for Underprivileged Children

Context of the Intervention. This project develops a value-focused Decision Analysis intervention designed and deployed to support the Csányi Foundation (Budapest, Hungary) in selecting underprivileged children to enter the Foundation's Educational program.

The Csányi Foundation is a non-profit organization that was set up in Hungary to help provide gifted children from disadvantaged backgrounds with the means to help nurture and develop their talent.

Over the last few years the Foundation has developed and grown with the help of prominent specialists and with the increasing financial support that has been given by people in Hungary.

At present, the Foundation helps over 350 children and young people, 56 of whom are already at university, with nine participants having completed the Foundation's career program. The goal is for every participant to have university qualifications, and for one thousand children to be involved in the program. Mentors and teachers help children to develop their talents over a period of 12 to 14 years, starting from the fifth year in elementary school. From 2017, 380 children and their families will receive grants through the Career Program.

The core focus of the Foundation is its Talent Development Program. The primary purpose of the program is to support children in achieving each milestone of their life, by overcoming socio-economic disadvantages. Every child thus benefits from nine months of lessons every year, tailored to their individualized plans, that include mathematics, creative writing, information technology, foreign languages, drama and self-awareness.

To achieve its mission, the foundation announces every year a tender for 15 applicants to enter the Foundation's Educational program. Gifted children from underprivileged backgrounds who are completing their fourth year in primary school are eligible to apply to the program.

As the Foundation spends approximately 4700 USD/year/child (operating costs excluded), a crucial objective of the organization is to minimize dropout rates. However, mentors have highlighted that a large number of children who were admitted into the Foundation do not fit into the profile and the focus of the program for disparate reasons. Consequently, more than 25% of the children quit the program too early.

After thoroughly examining these cases, the board of directors has concluded that these problems are caused by the lack of an adequate selection process. The board also acknowledged that the selection criteria are not entirely reflective of the objectives of the Foundation and that the fundamental objectives have not been thoroughly discussed.

The Foundation's board thus set out the goal to improve the selection process for the admissions season in May 2017[3].

The selection of underprivileged children for educational programs represents an important decision-making problem as, if successful, it can help children to realize the positive attitude and behavioral changes that will help them become happy, healthy, and successful adults, thus contributing to decreased unemployment rates, improved quality of life and better social integration. Currently, the Foundation relies on multiple criteria for the selection of children to enter the program, such as the children's socio-economic status, cooperation skills and talents.

Moreover, the selection of underprivileged children to enter the Csányi Foundation program is important to many stakeholders, including: the children and their families, the founder, the donors, the local government and policy makers, charitable NGOs, educational institutions, local community centers and other Foundations.

This decision-making problem is not only important but also more complex than general personnel selection processes, as it is challenged by the need for increased sensitivity and the presence of emotional factors that can increase subjectivity in the process and disagreement among decision makers.

Results. Given the need to improve the children selection process by better identifying the selection criteria that will allow the Foundation to meet its objectives and by

---

[3] This project has been awarded the INFORMS Decision Analysis Practice Award 2017.

fostering consensus among the board members, this project developed a value focused decision analysis intervention.

The first phase of the intervention thus consisted in using the list of devices and key questions presented in Table 1 to help decision makers identify their objectives. In particular, we developed face to face structured interviews with two board members, i.e. the President of the Advisory Board and the Operational Director of the Foundation. The list of questions presented in Table 1 has thus been adapted to fit the decision-making problem under analysis (i.e. underprivileged children selection for educational programs) and has been organized as a questionnaire to support the discussion with the Board members.

Table 4 represents the result of the Value Focused Thinking stage. While some key concerns and objectives were part of the initial mental model of the decision makers, those objectives highlighted in grey in Table 4 represent the new objectives that were identified after discussing the answers to the Value Focused Thinking questions. This now comprehensive set of objectives became the input to a Multi Attribute Value Theory model [22]. This model allowed the Foundation to evaluate the performance of each candidate on all the considered objectives, to assess the relative importance of each objective through sound elicitation protocols and to obtain an overall ranking of suitable candidates to be welcomed in the Foundation activities from the year 2017.

Table 4. Results of the combined Value Focused Thinking and MCDA approach [19]

	Key concerns	Fundamental objectives identified	Weight
Identifying the most suitable candidates for the Foundation program	Cognitive abilities	Maximise mathematical abilities	14.04%
		Maximise reading comprehension	15.60%
	Socio-economic background	Maximise level of underprivileged background	17.33%
		Maximise level of support from the children's family	13.86%
	Social competences	Maximise communication skills	1.70%
		Maximise empathy	6.93%
		Maximise cooperative attitude	4.16%
		Maximise leadership ability	2.23%
		Maximise attention	0.90%
		Maximise creativity	2.77%
	Personality	Maximise open mindedness	4.85%
		Maximise character's positive features	5.40%
		Maximise emotional steadiness	6.07%
	Mentors' perception	Maximise mentors' approval	3.47%
		Maximise mentors' enthusiasm	0.70%

In particular, the weights shown in Table 4 were obtained through structured interviews with both the President of the Advisory Board and the Operational Director of the Foundation. In this structured interviews we used the bisection elicitation protocol [e.g. 17] to help the decision makers visualize the ranges of performance on each objective and think about trade-offs among them.

Lessons Learned. As shown in Table 4, the solution proposed in this intervention allowed to generate 40% more objectives with respect to the initial concerns that the Decision Makers had in mind.

Moreover, the newly identified objectives concerning the "level of family support", "emotional steadiness", "character" and "open-mindedness" were recognized as the 4th, 6th, 7th and 8th respectively most important objectives among the 15 identified ones. These findings confirm again that, when confronted with important decisions, we fail to think about nearly half of the relevant objectives that we later recognize as important and that the objectives that are missed are indeed among the most relevant ones.

This intervention and its findings supported the selection process for the academic year 2017/2018, when the Foundation opened applications in two cities in Hungary for a total of 17 children. The feedback provided by the President of the Foundation Advisory Board and the Foundation Operational Director highlighted that the questions that were more effective in helping them to identify new objectives were the questions about alternatives, about problems and shortcomings and about goals, constraints and guidelines. This last category of questions in particular was the one that allowed to identify the new objective about the level of family support, which emerged as a crucial indicator.

The key impacts observed after this behavioral decision analysis intervention can be summarized as follows:

(i) learning effect: the Foundation's members indeed discovered new important objectives for the program thanks to the combined value focused thinking and MCDA intervention;

(ii) consensus building: the facilitated modelling approach employed in the intervention allowed the board members to adopt an inclusive perspective throughout the process and thus achieve consensus on both objectives to be achieved and candidates to be selected;

(iii) innovation: after the intervention the Foundation board voted on the embedding of the new tool in all subsequent selection processes and unanimously approved it;

(iv) reduced drop-out rates: after one year monitoring of the Foundation program, no children have drop-out for the academic year 2017–2018.

4   Conclusions

Research has shown that failure to recognize relevant objectives could be viewed alongside off-cited perceptual and cognitive biases [e.g. 20] as a fundamental cause of decision shortcomings. Indeed, Bond et al. [6] suggest that choice models based on dimensions an agent has reported to be important will frequently suffer from omitted variable bias, because the agent will have overlooked several factors of genuine importance to the decision.

To the extent that contemplation of one's objectives involves a substantial amount of reasoned deliberation (alignment of personal values to attributes of the decision, consideration of alternative outcomes, logical thinking etc.), narrow and shallow thinking may be viewed as deficiencies in system 2 processing, and interventions designed to counter these deficiencies may do so by stimulating the activity of this system [5].

The results and lessons learned discussed in this paper confirm key findings from the experimental literature on human limitations with respect to objectives' generation within decision making processes. However, this paper has an added value as it proposes two field studies, with real decision makers within their decision-making environment (public and private, respectively), as opposed to laboratory experiments.

The final key messages from this paper can be summarized as follows:

(i) Using the ten categories of value focused thinking devices allows to improve the decision frame, which is probably the most crucial component of the whole decision-making process. The different questions illustrated in Table 1 help indeed to move from a usually narrow and vague frame of the decision to a broader and more focused one. Both projects discussed in this paper showed that the list of objectives generated as a result of the collaborative value focused thinking approach leads to better quality alternatives.

(ii) Both projects also demonstrated that the ten categories of value focused thinking devices (Table 1) are effective in engaging system 2 and help to avoid the omitted variable bias. This seems an interesting finding compared to recent research on debiasing techniques' effectiveness for instance on overconfidence [21]. In a series of experiments on debiasing overprecision Ferretti et al. [21] indeed showed that think harder strategies (e.g. use of counterfactuals or hypothetical betting questions) are less effective compared to automatic rules (i.e. automatic stretching of the provided intervals). In both projects discussed in this paper the efficacy of the value focused thinking devices was demonstrated both in terms of quantity of objectives (roughly 40% more in both interventions) and quality of the same objectives (with the newly generated objectives consistently being among the most important ones).

Following from the previous consideration, an interesting direction for future research may explore and compare the efficacy of different approaches to support the generation of objectives, for instance stakeholders' analysis, cognitive mapping, SWOT analysis and value focused thinking devices.

In conclusion, this paper contributes to the debate on nudging versus boosting approaches [e.g. 10] as it shows how a process boost (i.e. thought-provoking questions based on the value focused thinking approach) can improve subjects' active competences for better and more empowered strategic decisions, as well as generate new versatile capabilities. For both projects the involved organizations provided indeed a very positive feedback on the developed collaborative process and expressed commitment for the embedding of the discovered approach in all future strategic decisions of the organization.

References

1. Keeney, R.L.: Value-Focused Thinking: A Path to Creative Decision Making. Harvard University Press, Cambridge (1992)
2. Spetzler, C., Winter, H., Meyer, J.: Decision Quality: Value Creation from Better Business Decisions. Wiley, Hoboken (2016)
3. Franklin, B.: Letter to Joseph Priestly (1772). (Reprinted in Isaacson, W. (ed.): A Benjamin Franklin Reader. Simon & Schuster, New York (2005)
4. Keeney, R.L., Raiffa, H.: Decisions with Multiple Objectives: Preferences and Value Tradeoffs. Wiley, New York (1976). (Reprinted by Cambridge University Press (1993)
5. Bond, S.D., Carlson, K.A., Keeney, R.L.: Improving the generation of decision objectives. Decis. Anal. **7**(3), 238–255 (2010)
6. Bond, S.D., Carlson, K.A., Keeney, R.L.: Generating objectives: can decision makers articulate what they want? Manag. Sci. **54**(1), 56–70 (2008)
7. Trope, Y., Liberman, N.: Temporal construal and time-dependent changes in preference. J. Pers. Soc. Psychol. **79**(6), 876–889 (2000)
8. Tulving, E., Pearlstone, Z.: Availability versus accessibility of information in memory for words. J. Verbal Learn. Verbal Behav. **5**(4), 381–391 (1966)
9. Keeney, R.L.: Value-focused thinking: identifying decision opportunities and creating alternatives. Eur. J. Oper. Res. **92**(3), 537–549 (1996)
10. Hertwig, R., Grüne-Yanoff, T.: Nudging and boosting: steering or empowering good decisions. Perspect. Psychol. Sci. **12**, 973–986 (2017)
11. Parnell, S.G., Bresnick, T.A., Johnson, E.R.: Craft the decision objectives and value measures. In: Parnell, G.S., Bresnick, T.A., Tani, S.N., Johnson, E.R. (eds.) Handbook of Decision Analysis, 1st edn. Wiley, Hoboken (2013)
12. Butler, J.C., Dyer, J.S., Jia, J.: Using attributes to predict objectives in preference models. Decis. Anal. **3**(2), 100–116 (2006)
13. Belton, V., Stewart, T.J.: Multiple Criteria Decision Analysis: An Integrated Approach. Kluwer Academic Publishers, Boston (2002)
14. Kpoumié, A., Damart, S., Tsoukiàs, A.: Integrating cognitive mapping analysis into multi-criteria decision aiding. Cahier du Lamsade **322** (2013). https://hal.archives-ouvertes.fr/hal-00875480
15. Stewart, T.J., Joubert, A., Janssen, R.: MCDA framework for fishing rights allocation in South Africa. Group Decis. Negot. **19**, 247–265 (2010)
16. Ferretti, V., Gandino, E.: Co-designing the solution space for rural regeneration in a new World Heritage site: a choice experiments approach. Eur. J. Oper. Res. **268**(3), 1077–1091 (2018)
17. Ferretti, V.: From stakeholders analysis to cognitive mapping and multi attribute value theory: an integrated approach for policy support. Eur. J. Oper. Res. **253**(2), 524–541 (2016)
18. Ferretti, V., Degioanni, A.: How to support the design and evaluation of redevelopment projects for disused railways? A methodological proposal and key lessons learned. Transp. Res. Part D **52**, 29–48 (2017)
19. Ferretti, V.: Framing, expanding and visualising: lessons learned from behavioural decision analysis in action. In: Viale, R. (ed.) Routledge Handbook of Bounded Rationality (under review)

20. Tversky, A., Kahneman, D.: The framing of decisions and the psychology of choice. Science **211**(4481), 453–458 (1981)
21. Ferretti, V., Montibeller, G., von Winterfeldt, D.: Testing the effectiveness of de-biasing techniques to reduce overprecision. Manag. Sci. (under review)
22. Keeney, R.L., Raiffa, H.: Decisions with Multiple Objectives: Preferences and Value Trade-Offs. Wiley, New York (1976)

Identifying and Ranking Critical Success Factors for Implementing Financial Education in Taiwan Elementary Schools

Ching Chih Tseng$^{(\boxtimes)}$ⓘ and Tzu Ning Kouⓘ

Dayeh University, Changhua 51591, Taiwan
cctseng@mail.dyu.edu.tw

Abstract. The dual card storm which occurred ten more years ago is still fresh in Taiwan's memory. To improve the financial literacy of children, financial education (FE) is mandated to include into primary school curriculum from 2011 in Taiwan. Research, addressing various issues of FE has appeared recently, but far fewer focused on identifying critical success factors (CSFs) for implementing FE. Therefore, this study is one of the first attempts to gain an understanding of the CSFs in this issue. This research is firstly based on financial literacy education discussed in relevant literature and conducts a two round Delphi survey to identify the CSFs. And then, the study ranks the CSFs using analytic hierarchy process (AHP) method. Our research reveals that, for successfully implement FE, 14 CSFs are identified. Finally, the research gives some discussion on the findings and, furthermore, presents two main theoretical implications as well as contributes two main managerial implications.

Keywords: Financial education · Critical success factors · Delphi method · Analytic hierarchy process

1 Introduction

Fueled by rising consumption among children and youth and changing perspectives on the economic lives of children, there has been a groundswell of interest in financial education aimed at young people (Greenspan 2005; Sherraden et al. 2011). However, the low level of teen financial literacy has been documented by various surveys conducted by Jump$tart and other organizations (Varcoe et al. 2005). These findings also hold true in Taiwan. The dual card (i.e., credit card and cash card) storm occurred in 2006 is still fresh in Taiwan's memory. Even after large relief expenditures, a large number of card slaves still bear the consequences of bad decisions in that year. One reason why they became card slaves is they were not financially literate—a factor that is causing great concern among government officials and educators (Card International 2007).

Many experts and educators proposed that financial education will help prepare young people to make sound financial decisions in an increasingly complex economic environment (Hilgert et al. 2003; Lucey and Giannangelo 2006; Lusardi and Mitchell 2007; Sherraden et al. 2011). Thus, a consensus has emerged around the need to expand financial education in the developed and developing world. In schools, if we

D. C. Morais et al. (Eds.): GDN 2019, LNBIP 351, pp. 106–119, 2019.
https://doi.org/10.1007/978-3-030-21711-2_9

can educate students about finance to benefit their future and the decisions they make, then students will have a better understanding of how the real world work. In both developing and developed economies, the awareness of the importance of FE has gained momentum among policy makers leading notably to the development of an increasing number of tailored national strategies for financial education in recent years (Grifoni and Messy 2012).

When should FE start in the life of an individual? Some works (Boshara and Emmons 2015; Drever et al. 2015) highlighted that the foundations of financial knowledge are actually built during childhood. It have been showed that children are able to handle basic economic concepts well (Roos et al. 2005; Chan and McNeal 2006; Leiser and Halachmi 2006; Otto et al. 2006; Bucciol and Piovesan 2011; Hospido et al. 2015). These results supported the idea that young people should receive FE very early in their lives.

Including FE into elementary school curricula seems like one possible way to improve financial knowledge and has been an enormous step forward because prevention seems to be the most effective measure (Opletalová 2015). Lusardi and Mitchell (2014) pointed out that greater financial knowledge is also positively related to savvier saving and investment decisions, more retirement planning, greater participation in the stock market, and greater wealth accumulation. Younger students can learn financial topics and that learning is associated with improved attitudes and behaviors which, if sustained, may result in increased financial capability later in life (Batty et al. 2015).

In 2005, the Ministry of Education, Taiwan, initiated a program named as "Financial Literacy Promotion Programme" to cultivate financial literacy for elementary school students. Since 2011, FE has been mandated to include into elementary school curricula, in Taiwan. Based on past empirical studies, Taiwanese elementary and junior high school teachers had the problems of insufficient financial knowledge and few opportunities for on-the-job study and advancement (Deng et al. 2013). Thus, based on reasonable inference, the performance of FE implementation in Taiwan is not significant.

To enhance the effectiveness of the FE, the elements or factors that affect FE successful implementation should be figured out. Thus education authority can make more effort in higher priority factors to improve the performance. However, far fewer research addressed this issue. In practice, it is hard to improve all influencing factors simultaneously. A feasible method is to just focus on some most urgent and important identified factors, and to implement them. Therefore, the concept of critical success factors (CSFs) is adopted in this study. The objective of this study is to identify and rank the CSFs for implementing FE in elementary schools, thus, the education authority can just focus on the higher priority factors to greatly improve the effectiveness and performance.

To achieve the objective, this paper proposes a hybrid approach integrating Delphi method and analytic hierarchy process (AHP). Due to the absence of much useful literature, the Delphi method is used to identify CSFs. Then AHP is employed to rank the CSFs. There are two reasons we adopt the AHP technique. One, it is an easy-to-understand mean to reach consensus in a group decision context, and it is worthwhile to illustrate the application of this technique in a case study setting where one is confronted with subjective decisions and feelings (van de Water and de Vries 2006). The other is it complements the weakness of score ranking in the Delphi method which does not allow a participant to weigh the relative difference between item rankings (Couger 1988).

The rest of this paper is organized as follows. Section 2 presents a literature review; Sect. 3 develops the research framework of this study; the results and discussion are presented in Sect. 4; Conclusion and implications are provided in Sect. 5.

2   Literature Review

2.1   Financial Education

As early as 2005, the OECD Recommendation specially advised that "FE should start at school. People should be educated about financial matters as early as possible in their lives (OECD 2005). The OECD also defined the term FE as "FE is the process by which financial consumers/investors improve their understanding of financial products and concepts and, through information, instruction and/or objective advice, develop the skills and confidence to become more aware of financial risks and opportunities, to make informed choices, to know where to go for help, and to take other effective actions to improve their financial well-being "(Lusardi and Mitchell 2007). (Hospido et al. 2015) argued that two main reasons support this recommendation. One is younger generations are likely to face ever-increasing complexity in financial products, services and markets; the other is schools are well positioned to advance financial literacy among all demographic groups and reduce financial literacy gaps and inequalities.

Many studies supported the idea that young people should receive FE very early in their lives, a lot of works regarding programs focused on children (Gross et al. 2005; Carlin and Robinson 2012; Becchetti et al. 2013; Romagnoli and Trifilidis 2013; Alan and Ertac 2014; Batty et al. 2015; Lührmann et al. 2015; Migheli and Moscarola 2017) as well as attitudes and behaviors of teen-agers involved in the programs. Otto et al. (2005) found that children learn to manage money in a formal manner between the ages of 9 and 12. For six years old children, they had not yet experienced the applicative value of saving and had been saving money because their parents expected them to do so. On the other hand, older children had already been saving their money for the sake of saving and to avoid the temptation to spend it.

Some works (Schug 1987; Webley and de Nyhus 2006; Scheinholtz et al. 2011; Batty et al. 2017) on children's cognitive development and economic understanding indicated not only that children can understand financial concepts at younger ages than high school, but also that their understanding was well developed by age 12. Including financial capability into elementary and secondary school education programs seems like one possible way to prevent an unhealthy debt burden and to utilize and create financial reserves.

2.2   CSFs

The term success factors was first discussed by Daniel (1961) and CSFs were introduced by Rockart (1979) as a way to help senior executives define their information needs for the purpose of managing their organizations. Rockart (1979) defined CSFs as "the limited number of areas in business, if they are satisfactory, will ensure successful competitive performance for the organization". Islam (2010) described CSFs as "factors that must be implemented in order to successfully address the challenge". Meibodi and Monavvarian (2010) viewed CSF as readily evident or sometimes as an invisible set of activities that an organization must undertake in order to achieve its objectives. Ab Talib et al. (2015) defined CSFs as few factors that are critical for the success of a company and they must receive careful and constant attention from the managers. So far, there are a variety of definitions of CSFs. Definition of CSFs does not only concern a firm's success but also a firm's ability to overcome the challenges faced.

The original concept of CSFs was formulated as a consequence of revising Parteo's findings and stated that, for many events, roughly 80% of the effects come from 20% of the causes. In case of CSFs it means, that there is no need to examine all CSFs but rather to focus on 20% of them, because those are the factors that will in decisive manner (80%) be responsible for a success or a failure of the enterprise.

Research, addressing various issues on FE has appeared recently, but far fewer focused on identifying CSFs for implementing FE. Therefore, in this study literature review along with experts' opinions is used to identify CSFs for FE implementation. In the absence of much useful literature on CSFs of this issue, this study first identify the dimensions of CSFs from literature review and then adopts Delphi method to further identify their respective CSFs that are jointly agreed by a panel of FE experts and practitioners. The Delphi method was deemed to be the most appropriate method for this study because it allows the gathering of subjective judgements which are moderated through group consensus (Carson et al. 2001).

Through literature review, four dimensions: government policy, school and teacher, family function, and social resources support, are identified and described as follows.

Dimension of Government Policy. One of the possible critical success factors for implementing FE is government policy and action plan. A significant number of both emerging and developed countries seek to improve financial inclusion through their national strategy for FE (OECD 2015). Several national FE initiatives are under way, many spearheaded by Federal agencies (Fox et al. 2005).

Dimension of School and Teacher. FE in elementary schools and teachers are important to prepare children to deal with the complexities of today's financial world and promote financial literacy. Granville et al. (2009) pointed out that evaluations of best practice in the provision of FE identify the following six critical factors: allocation of curriculum time for FE, effective leadership and co-ordination from policy makers and senior management within schools, teachers being confident in their knowledge and ability, FE being engaging for students, teachers making effective use of the materials and resources, and appropriate systems for monitoring pupil progress – attainment of key learning outcomes. These findings showed that the importance of the roles of school and teacher on FE.

Dimension of Family Function. When children were young, the family is a primary socialization unit for them and serves as a filtering unit for information from the outside. By the time children reach school, they had the foundations of their values, beliefs, attitudes, expectations, and motivations about money already established (Schug 1987). But, children learn by observation, practice, and intentional teaching. Many of their values, beliefs and attitudes about money were established in family through observation and example (Danes 1994).

Dimension of Social Resources Support. A great number of public and private entities engage in personal FE (Fox et al. 2005). Providers of FE programs from the Fannie Mae report included (1) community organizations, (2) Cooperative Extension Service, (3) businesses, (4) faith-based organizations, (5) community colleges, and (6) the U.S. Military (Vitt 2001). Moreover, from the US to the 27 members of the EU, governments, central banks and other primary financial institutions and authorities have designed and implemented programs of financial literacy targeted to elementary and secondary schools (Migheli and Moscarola 2017). Besides, the effort from social resource such as cooperate, bank, community, etc., is necessary. This shows that support from social resource support plays an important role.

Additionally, McCormick (2009) conducted a literature review on the current state of youth FE and policy. Five key factors and promising practices for FE effectiveness were summarized as follows.

- Youth FE must permeate the entire K-12 setting rather than wait until the middle or high school years for introduction.
- It must demonstrate relevance to students in order to engage their motivation.
- Beyond teaching students to handle their cash, it must be designed to forge understandings of the relationships among money, work, investments, credit, bill payment, retirement planning, taxes, and so forth.
- Systemically, it must be mandated by state academic standards in order to gain widespread implementation and time and resource commitments from teachers and school systems.
- Teacher training and professional development opportunities are a necessary corollary to successful program implementation.

3   Research Methodology

Our research is split into two stages. In the absence of much useful literature on CSFs for a successful FE implementation, the first stage employed the Delphi method to identify the CSFs. In the second stage, the AHP method was applied to rank the identified CSFs and their dimensions from the experts' point of view.

3.1   Delphi Method

The Delphi method is a pragmatic research method created in the 1950s by researchers at the RAND Corporation for use in policy making, organizational decision making, and to inform direct practices (Brady 2015). It is an opinion survey technique that gathers information from experts' opinions. Identifying and selecting CSFs from an experts' point of view can be performed using the Delphi method (Duncan 1995).

The Delphi method in this study incorporates a group survey technique involving an iterative multistage process that is widely applied in social sciences, information system, information technology, and health fields (Hasson et al. 2000; Skulmoski et al. 2007), as well as in public policy design and implementation (Alder and Ziglio 1996), education (Foth et al. 2016).

In this study, a Delphi panel composes of fourteen Taiwanese FE experts was established. These experts include four scholars from university, one from Financial Supervisory Commission (FSC), an independent government agency, two from private institution practitioners, two from the Ministry of Education, two from the Ministry of Finance, and three elementary school teachers who teach FE. Thus, the size of such a Delphi panel is deemed suitably representative. The panelists have all been substantially involved in the FE research or practices many years (at least five years). Alder and Ziglio 1996) asserted that useful results can be obtained from small group of 10–15 experts. Beyond this number, further increases in understandings are small and not worth the cost or the time spent in additional interview (Carson et al. 2001). Moreover, this research assumes that experts' opinion can be of significant value in situations where knowledge or theory is incomplete, as in our case.

The Delphi survey in this study comprises two rounds. During the first round the authors conducted face-to-face interviews with each panelist (and phone interviews in some cases due to geographical constraints), and these varied in duration from one to one-and-half hours. Rather than having an open-end question, the authors employed a different approach from traditional Delphi methods by beginning with a list of dimensions derived from comprehensive literature reviews. Having a prior theory has advantages such as allowing the opening and probe questions to be more direct and effective, and helping the researcher recognize when something important has been said (Carson et al. 2001). However, it seems to be far few literature with respect to CSFs for a successful FE implementation. At the beginning of the interviews it was explained that the study focused on identifying the CSFs for implementing FE and that based on preliminary literature review, there are four identified major dimensions. Also, the panelist is asked to mention or write down any CSFs or other dimensions and then further is probed questions would follow in order to gather more details on those factors. The panelists were indeed encouraged to suggest any factors that they deemed critical. After the interview, further clarifications (if any) were made by follow-up phone calls and email communications.

In what follows, the suggested factors were recorded, clarified, and consolidated them into a single list. The list contained four dimensions and 19 CSFs.

In the second round, the list was distributed among the participants to facilitate comparison of the panel's conceptual differences. However, none of them nominated any additional factors of their own. In addition, the panel confirmed that the classification of dimensions is appropriate. Also, based on feedback from panel members some further minor changes were incorporated. For example, several factors are grouped together because of the closed interrelationship. Finally, we obtain four dimensions and obtain totally 14 CSFs (form original 19 CSFs) for implementing FE in Taiwanese elementary schools as listed in Table 1.

Table 1. CSFs for implementing FE

Level 1 dimension	Level 2 critical factors
Government policy and action	(A1) A holistic and coherent FE program
	(A2) FE as an important national education policy
	(A3) Provide sufficient funds for FE implementation
	(A4) Create advisory and counseling teams
	(A5) Establish FE resource centers
School teaching and advocacy activities	(B1) Prepare FE teaching materials and aids
	(B2) Teachers with financial professional knowledge and good FE teaching strategy
	(B3) Strengthen FE advocacy activities
	(B4) Teachers recognize and earnestly implement FE
Family function	(C1) Parent with good financial literacy
	(C2) Help your child build right money concept and consumption skills
	(C3) Make good use of FE resources and good parent-child interaction
Social resource support	(D1) Encourage private enterprises and organizations to support FE
	(D2) Promote financial experts to provide guidance and teaching advice

3.2   AHP Approach

AHP is well established methodology that was developed by Saaty in 1977. Saaty (1980) defined AHP as a hierarchical decomposing decision method for a complex multi-criteria decision problem. AHP is widely used to show the importance or weights of the factors (Zahedi 1986). AHP is extensively used for solving different multi-criteria decision making (MCDM) problems and is widely applied in the various fields such as marketing (Chen and Wang 2010), technology transfer adoption (Lee et al. 2012), supply chain management (Govindan et al. 2014), technology transfer (Kumar et al. 2015), energy technologies adoption (Luthra et al. 2015), supplier selection problem (Dweiri et al. 2016), and education (Hsueh and Su 2016; Thanassoulis et al. 2017). Also, Anis and Islam (2015) conducted a literature review of the AHP applied in higher-learning institutions.

AHP technique is simple, systematic, scientific, dependable, and user friendly at the same time because of availability of suitable software (i.e., EXPERT CHOICE) to calculate priority matrices from comparison matrices. A decision-maker should determine the weights by conducting pair-wise comparisons between various criteria. For these reasons, we propose the AHP to rank the identified CSFs. Chen and Wang (2010) pointed out that the main procedures of AHP are: (1) determine the objective and the attributes of evaluation; (2) develop hierarchical structure levels with goals, contracture, criteria and the alternatives; (3) find out the importance of different attributes with respect to the goals.

A Hierarchic Framework. According to the AHP steps as above, the relevant literature on FE are firstly reviewed in Sect. 2, and then several CSFs were collated by conducting a two-round Delphi surveys in Sect. 3.1. Finally, we obtain four dimensions and totally 14 CSFs as listed in Table 1. Thus, these dimensions and CSFs had considerable degree of content validity.

Pairwise Comparison Matrix. Further details for AHP process are as follows (Saaty 1990, 1994; Saaty and Vargas 2000; Chen and Wang 2010):

Step 1. Constructing a pair-wise comparison matrix with a scale of relative importance. An attribute compared with itself is always attributed to value 1, so all the main diagonal entries of the pair-wise comparison matrix are 1. Numbers 3, 5, 7, and 9 mean moderate importance, strong importance, very important, and absolutely important; and 2, 4, 6, and 8 for compromise between 3, 5, 7, 9.

Step 2. Finding the relative normalized weight (wj) of each attribute by calculating the geometric mean (GM) of the ith row normalizes the geometric means of rows in the comparison matrix. The geometric mean method of AHP is used to find out the relative normalized weights of the attributes because of its simplicity and ease to find out the maximum eigenvalue and reduce the inconsistency in judgments.

Step 3. Finding out the maximum eigenvalue λ_{max} (Saaty 1994).

Step 4. Calculating the consistency index as equation $CI = (\lambda_{max} - m)/(m - 1)$. Here, m denotes that the number of CSFs. The consistency in the judgments of relative importance of attributes reflects the cognition of the analyst.

Step 5. Obtaining the random index (RI) for the number of attributes used in decision-making.

Step 6. Calculating the consistency ratio $CR = CI/RI$. Usually, a CR of 0.1 or less is considered as acceptable and reflects an informed judgment that could be attributed to the knowledge of the analyst.

4   Results and Discussion

This research sent out 25 AHP questionnaires and all of the respondents (five form university scholars, two from FSC, four from the Ministry of Finance, four from the Ministry of Education, ten from first line staffs and teachers in elementary schools) have been involved with the field FE over 5 years.

4.1   Results

According to the data from the questionnaire, we figured out the weight of each item by EXPERT CHOICE 2000. After computing, we found that nearly all replies to the questionnaire reached a consistency ratio (i.e. $CR = 0.0054$) of less than 0.1, hence the decision maker's pair-wise comparison matrices are acceptable. The overall weight of the dimensions and CSFs, and its ranking is shown in Table 2. As the results in Table 2 show, the ranking of the weights of the dimensions is: family function (0.470), school teaching and advocacy activities (0.225), government policy and action (0.217), and social resource support (0.088). These results manifest the most influential dimension is family function, and the least influential one is social resource support.

On the other hand, there are 14 CSFs are shown as Table 2. The top ten factors and their weights of global ranking are: (C1) "parents with good financial literacy" (0.2747), (C2) "help your child build right money concept and consumption skills" (0.1405), (B4) "teachers recognize and earnestly implement FE" (0.1405), (B2) "teachers with financial professional knowledge and good FE teaching strategy" (0.0745), (A2) "FE as an important national education policy" (0.0741), (A1) "a holistic and coherent FE guideline" (0.0632), (C3) "make good use of FE resources and good parent-child interaction" (0.0548), (D1) "facilitate private enterprises and organizations to support FE" (0.0536), (A3) "provide sufficient funds for FE implementation" (0.0415) and (B1) "prepare FE teaching materials and aids" (0.0345). These results indicate that the factors are fully distributed over all dimensions, and this distribution is more in accordance with dimensions ranking results.

Table 2. The AHP weight and ranking of dimension and critical success factor

Dimension	Weight	Ranking	CSFs	Weight (global)	Ranking (local)	Ranking (global)
Government policy and action	0.217	3	A1. A holistic and coherent FE guideline	0.0632	2	6
			A2. FE as an important national education policy	0.0741	1	5
			A3. Provide sufficient funds for FE implementation	0.0415	3	9
			A4. Create advisory and counseling teams	0.0174	5	14
			A5. Establish FE resource centers	0.0209	4	12
School teaching and advocacy activities	0.225	2	B1. Prepare FE teaching materials and aids	0.0345	3	10
			B2. Teachers with financial professional knowledge and good FE teaching strategy	0.0745	2	4
			B3. Strengthen FE advocacy activities	0.0176	4	13
			B4. Teachers recognize and earnestly implement FE	0.0980	1	3
Family function	0.470	1	C1. Parent with good financial literacy	0.2747	1	1
			C2. Help your child build right money concept and consumption skills	0.1405	2	2
			C3. Make good use of FE resources and good parent-child interaction	0.0548	3	7

(continued)

Table 2. (*continued*)

Dimension	Weight	Ranking	CSFs	Weight (global)	Ranking (local)	Ranking (global)
Social resource supporting system	0.088	4	D1. Facilitate private enterprises and organizations to support FE	0.0536	1	8
			D2. Promote financial experts to provide guidance and teaching advice	0.0343	2	11

C.I = 0.0049 < 0.1; C.R = 0.0054 < 0.1

4.2 Discussion

From global ranking shown in Table 2, some findings are described as follows.

1. Parents play a key role in FE implementation and are the first teachers in child's growth and development process. The top two CSFs are located in dimension of family function, with a total percentage of 41.5%, far larger than other CSFs. This result indicates that the government should give more focus on dimension (C) "family function", especially on critical factors of (C1) "parents with good financial literacy" and (C2) "help their children build right money concept and consumption skills". Parents with good financial literacy can lead by example and teach their children right money concept and build good parent-child interaction, and then the children become a model of their next generation.
2. In school teaching and advocacy activities dimension, there are two CSFs in the third and fourth rankings. It shows that the first-line teachers are also important influencers in child's growth and development process. Teachers with good financial profession knowledge and with good teaching strategy who recognize and earnestly implement FE will enable FE be more effective in elementary schools.
3. The last three least influential CSFs are (A5) "establish a FE resource center", (B3) "strengthen FE advocacy activities", and (A4) "create advisory and counseling teams". These three CSFs with global weight values less than 0.03 are relatively less important.

However, the last three factors also cannot be neglected. Factor (A5) "establish a FE resource centers" can develop FE resources and tools to provide the first line educators to improve the learning interest of the students of elementary schools. Factor (B3) "strengthen FE advocacy activities" can establish an annual activity, for example, financial literacy month. In this annual activity, various types of FE activities or competitions can be held, and through these activities, it can really help teachers to implant financial literacy into the students. Currently, FE is not yet included in the formal curriculum in Taiwanese elementary school, the financial literacy education can only be integrated into other curriculums, so a simple and holistic content of financial literacy is difficult to provide.

As for factor (A4) "create advisory and counseling teams", the team are created and composed of members from government sectors such as the center bank of Taiwan, FSC, the Ministry of Finance, and the Ministry of Education. The objective of the teams are to train the first line teachers as the seeds to help achieve successful FE implementation.

However, many other important issues such as human rights are obliged to be integrated into elementary school curriculum. Although the education authority believes that the normal subjects included these issues do not add too much burden to the elementary school teachers, in reality, the result is the opposite. Too much burden may reduce the willingness to recognize and earnestly implement FE.

5   Conclusions and Implications

Identifying and ranking the CSFs for successfully implementing FE is a complex issue. This research has identified and ranked 14 CSFs. Based on the identified the CSFs and their priorities, the government sectors can strategically allocate resources to the higher CFSs so as to improve the FE implementation performance. By using these priorities, government sectors can also decide which CSFs they will focus on first, next, and then last.

After conducting a comprehensive review of relevant literature and a two-round Delphi survey, four dimensions and 14 CSFs were identified. Thereafter, the study ranked the weights of the CSFs using the AHP method. (C1) "parents with good financial literacy", (C2) "help your child build right money concept and consumption skills", (B4) "teachers recognize and earnestly implement FE", (B2) "teachers with financial professional knowledge and good FE teaching strategy", and (A2) "FE as an important national education policy" are found to be top five CSFs.

This research presents two main theoretical implications. First, it contributes to the literature on the issue of identifying the CSFs influencing FE successful implementation. The obtained theoretical CSFs could be of potential value to future researchers in FE. Second, this research contains an approach to identify and prioritize the ranking of CSFs by a hybrid methodology combing the Delphi method and the AHP approach. A theoretical framework is proposed.

Moreover, the research also contributes two main managerial implications. First, based on the findings, the government strongly promote FE in elementary schools while not neglecting the important role of family function. This means that children learns good financial literacy, not only from schools but also more from their parents. When a child grows up and becomes parent, he/she will also contribute to cultivate good financial literacy for the next generation. Second, the obtained priorities help the government and policy makers understand the relative importance of the CSFs. This is helpful to establish their strategic plans as they may not have sufficient resources to deal with all the factors simultaneously.

References

Ab Talib, M.S., Abdul Hamid, A.B., Zulfakar, M.H.: Halal supply chain critical success factors: a literature review. J. Islamic Mark. **6**(1), 44–71 (2015)

Aghaei Meibodi, L., Monavvarian, A.: Recognizing critical success factors to achieve the strategic goals of SAIPA Press. Bus Strategy Ser. **11**(2), 124–133 (2010)

Adler, M., Ziglio, E.: Gazing into the Oracle: The Delphi Method and Its Application to Social Policy and Public Health. Jessica Kingsley Publishers, London (1996)

Anis, A., Islam, R.: The application of analytic hierarchy process in higher-learning institutions: a literature review. J. Int. Bus. Entrep. Dev. 8(2), 166–182 (2015)

Batty, M., Collins, J.M., Odders-White, E.: Experimental evidence on the effects of financial education on elementary school students' knowledge, behavior, and attitudes. J. Consum. Aff. 49(1), 69–96 (2015)

Batty, M., Collins, M., O'rourke, C., Elizabeth, O.: Experiential Financial Literacy: A Field Study of My Classroom Economy (2017).https://www.ssc.wisc.edu/~jmcollin/wp/wp-content/uploads/2015/05/Boulder_MCE.pdf

Becchetti, L., Caiazza, S., Coviello, D.: Financial education and investment attitudes in high schools: evidence from a randomized experiment. Appl. Financ. Econ. 23(10), 817–836 (2013)

Boshara, R., Emmons, W.R.: A balance sheet perspective on financial success: why starting early matters. J. Cons. Aff. 49(1), 267–298 (2015)

Brady, S.R.: Utilizing and adapting the Delphi method for use in qualitative research. Int. J. Qual. Methods (2015). https://doi.org/10.1177/1609406915621381

Bucciol, A., Houser, D., Piovesan, M.: Temptation and productivity: a field experiment with children. J. Econ. Behav. Organ. 78(1–2), 126–136 (2011)

Bucciol, A., Piovesan, M.: Luck or cheating? A field experiment on honesty with children. J. Econ. Psychol. 32(1), 73–78 (2011)

Card International: Taiwan's credit crisis: the calm after the storm (2007). http://www.vrl-financial-news.com/cards–payments/cards-international/issues/ci-2007/ci375/taiwan%E2%80%99s-credit-crisis-the-ca.aspx

Carlin, B.I., Robinson, D.T.: What does financial literacy training teach us? J. Econ. Educ. 43(3), 235–247 (2012)

Carson, D., Gilmore, A., Perry, C., Gronhaug, K.: Qualitative Marketing Research. Sage, London (2001)

Chan, K., McNeal, J.U.: Chinese children's understanding of commercial communications: a comparison of cognitive development and social learning models. J. Econ. Psychol. 27(1), 36–56 (2006)

Chen, M.K., Wang, S.C.: The critical factors of success for information service industry in developing international market: using analytic hierarchy process (AHP) approach. Expert Syst. Appl. 37(1), 694–704 (2010)

Couger, J.D.: Key human resource issues in IS in the 1990s: views of IS executives versus human resource executives. Inf. Manag. 14(4), 161–174 (1988)

Daniel, D.R.: Management information crisis. Harvard Bus. Rev. 39, 111–121 (1961)

Deng, H.T., Chi, L.C., Teng, N.Y., Tang, T.C., Chen, C.L.: Influence of financial literacy of teachers on financial education teaching in elementary schools. Int. J. e-Educ. e-Bus. e-Manag. e-Learn. 3(1), 68–73 (2013)

Drever, A.I., Odders-White, E., Kalish, C.W., Else-Quest, N.M., Hoagland, E.M., Nelms, E.N.: Foundations of financial well-being: insights into the role of executive function, financial socialization, and experience-based learning in childhood and youth. J. Consum. Aff. 49(1), 13–38 (2015)

Duncan, N.B.: Capturing flexibility of information technology infrastructure: a study of resource characteristics and their measure. J. Manag. Inf. Syst. 12(2), 37–57 (1995)

Dweiri, F., Kumar, S., Khan, S.A., Jain, V.: Designing an integrated AHP based decision support system for supplier selection in automotive industry. Expert Syst. Appl. 62, 273–283 (2016)

Foth, T., et al.: The use of Delphi and Nominal Group Technique in nursing education: a review. Int. J. Nurs. Stud. 60, 112–120 (2016)

Fox, J., Bartholomae, S., Lee, J.: Building the case for financial education. J. Consum. Aff. **39**(1), 195–214 (2005)

Garcia, N., Grifoni, A., Lopez, J.C., Mejía, D.: Financial education in Latin America and the Caribbean: rationale, overview and way forward. In: OECD Working Papers on Finance, Insurance and Private Pensions, no. 33. OECD Publishing (2013). http://www.oecd.org/finance/financial-education/wp33finedulac.pdf

Govindan, K., Kaliyan, M., Kannan, D., Haq, A.N.: Barriers analysis for green supply chain management implementation in Indian industries using analytic hierarchy process. Int. J. Prod. Econ. **147**, 555–568 (2014)

Granville, S., Primrose, D., Chapman, M.: Evaluation of financial education in Scottish primary and secondary schools. Scottish Gov. Soc. Res. (2009). http://dera.ioe.ac.uk/379/1/0077311.pdf

Greenspan, A.: The importance of financial education today. Soc. Educ. **69**(2), 64–66 (2005)

Grifoni, A., Messy, F.: Current status of national strategies for financial education: a comparative analysis and relevant practices. In: OECD Working Papers on Finance, Insurance and Private Pensions, no. 16. OECD Publishing (2012). https://doi.org/10.1787/5k9bcwct7xmn-en

Gross, K., Ingham, J., Matasar, R.: Strong palliative, but not a panacea: results of an experiment teaching students about financial literacy. J. Student Financ. Aid **35**(2), 7–26 (2005)

Hasson, F., Keeney, S., McKenna, H.: Research guidelines for the Delphi survey technique. J. Adv. Nurs. **32**(4), 1008–1015 (2000)

Hilgert, M.A., Hogarth, J.M., Beverly, S.G.: Household financial management: the connection between knowledge and behavior. Fed. Reserve Bull. **89**, 309–322 (2003)

Hospido, L., Villanueva, E., Zamarro, G.: Finance for all: the impact of financial literacy training in compulsory secondary education in Spain (2015). https://www.econstor.eu/bitstream/10419/110143/1/dp8902.pdf

Hsueh, S.L., Su, F.L.: Critical factors that influence the success of cultivating seed teachers in environmental education. Eurasia J. Math. Sci. Tech. Educ. **12**(11) (2016). https://doi.org/10.12973/eurasia.2016.02306a

Kumar, S., Luthra, S., Haleem, A.: Benchmarking supply chains by analyzing technology transfer critical barriers using AHP approach. Benchmarking Int. J. **22**(4), 538–558 (2015)

Islam, R.: Critical success factors of the nine challenges in Malaysia's vision 2020. Socio-Econ. Plan. Sci. **44**(4), 199–211 (2010)

Lee, S., Kim, W., Kim, Y.M., Oh, K.J.: Using AHP to determine intangible priority factors for technology transfer adoption. Expert Syst. Appl. **39**(7), 6388–6395 (2012)

Leiser, D., Halachmi, R.B.: Children's understanding of market forces. J. Econ. Psychol. **27**(1), 6–19 (2006)

Lucey, T.A., Giannangelo, D.M.: Short changed: the importance of facilitating equitable financial education in urban society. Educ. Urban Soc. **38**(3), 268–287 (2006)

Lührmann, M., Serra-Garcia, M., Winter, J.: Teaching teenagers in finance: does it work? J. Bank. Financ. **54**, 160–174 (2015)

Lusardi, A., Mitchelli, O.S.: Financial literacy and retirement preparedness: evidence and implications for financial education. Bus. Econ. **42**(1), 35–44 (2007)

Lusardi, A., Mitchell, O.S.: The economic importance of financial literacy: theory and evidence. J. Econ. Lit. **52**(1), 5–44 (2014)

Lusardi, A., Mitchell, O.S., Curto, V.: Financial literacy among the young. J. Consum. Aff. **44**(2), 358–380 (2010)

Luthra, S., Kumar, S., Garg, D., Haleem, A.: Barriers to renewable/sustainable energy technologies adoption: Indian perspective. Renew. Sustain. Energy Rev. **41**, 762–776 (2015)

McCormick, M.: The effectiveness of youth financial education: a review of the literature (2009). https://files.eric.ed.gov/fulltext/ED540835.pdf

Migheli, M., Moscarola, F.C.: Gender differences in financial education: evidence from primary school. De Economist **165**(3), 321–347 (2017)

OECD Publishing: Improving financial literacy: analysis of issues and policies. Organization for Economic Co-operation and Development (2005). http://www.oecd.org/daf/fin/financial-education/improvingfinancial-literacyanalysisofissuesandpolicies.htm

Opletalová, A.: Financial education and financial literacy in the Czech education system. Procedia Soc. Behav. Sci. **171**, 1176–1184 (2015)

Otto, A.M., Schots, P.A., Westerman, J.A., Webley, P.: Children's use of saving strategies: an experimental approach. J. Econ. Psychol. **27**(1), 57–72 (2006)

Rockart, J.F.: Chief executives define their own data needs. Harvard Bus. Rev. **57**(2), 81–93 (1979)

Romagnoli, A., Trifilidis, M.: Does financial education at school work? Evidence from Italy. Bank of Italy, Economic Research and International Relations Area (2013). http://www.bancaditalia.it/pubblicazioni/qef/2013-0155/QEF_155.pdf

Roos, V., et al.: Money adventures: introducing economic concepts to preschool children in the South African context. J. Econ. Psychol. **26**(2), 243–254 (2005)

Saaty, T.L.: How to make a decision: the analytic hierarchy process. Eur. J. Oper. Res. **48**(1), 9–26 (1990)

Saaty, T.L.: How to make a decision: the analytic hierarchy process. Interfaces **24**(1), 19–43 (1994)

Saaty, T.L.: The Analytic Hierarchy Process. McGraw-Hill Book Co., New York (1980)

Scheinholtz, L., Holden, K., Kalish, C.: Cognitive development and children's understanding of personal finance. In: Lambdin, D. (ed.) Financial Decisions Across the Lifespan: Problems, Programs, and Prospects, pp. 29–47. Springer, New York (2011). https://doi.org/10.1007/978-1-4614-0475-0_3

Schug, M.C.: Children's understanding of economics. Elem. Sch. J. **87**(5), 507–518 (1987)

Sherraden, M.S., Johnson, L., Guo, B., Elliott, W.: Financial capability in children: effects of participation in a school-based financial education and savings program. J. Fam. Econ. Issues **32**(3), 385–399 (2011)

Skulmoski, G.J., Hartman, F.T., Krahn, J.: The Delphi method for graduate research. J. Inf. Technol. Educ. **6**, 1–21 (2007)

Thanassoulis, E., Dey, P.K., Petridis, K., Goniadis, I., Georgiou, A.C.: Evaluating higher education teaching performance using combined analytic hierarchy process and data envelopment analysis. J. Oper. Res. Soc. **68**(4), 431–445 (2017)

Varcoe, K.P., Martin, A., Devitto, Z., Go, C.: Using a financial education curriculum for teens. J. Financ. Couns. Plan. **16**(1), 63–71 (2005)

Vitt, L.A.: Personal finance and the rush to competence: financial literacy education in the US. A national field study commissioned and supported by the Fannie Mae Foundation. ISFS, Institute for Socio-Financial Studies (2001). https://www.isfs.org/documents-pdfs/rep-finliteracy.pdf

van de Water, H., de Vries, J.: Choosing a quality improvement project using the analytic hierarchy process. Int. J. Qual. Reliab. Manag. **23**(4), 409–425 (2006)

Webley, P., de Nyhus, E.K.: Parents' influence on children's future orientation and saving. J. Econ. Psychol. **27**(1), 140–164 (2006)

Conflict Resolution

War as a Technique of International Conflict Resolution – An Analytical Approach

Nadav Prawer◉ and John Zeleznikow(✉)◉

College of Business, Victoria University, Melbourne, Australia
Nadav.Prawer@gmail.com, John.Zeleznikow@vu.edu.au

Abstract. This paper considers the role of armed conflict, and attitudes to the study of armed conflict, in an evaluation of international conflict resolution. Its primary method is an investigation of the literature and consideration of analytical approaches adopted by International Conflict Resolution Research. We demonstrate that armed conflict is a major tool used by countries, with a high degree of effectiveness. We adopt a primarily theoretical, realist approach to analysis of the field.

Our findings and conclusions are that the exclusion of armed conflict from the analysis of international conflict resolution methods is not justified conceptually and leads to a warped analysis of party behaviours in conflict situations. We consider this as a substantial finding, which questions long-standing assumptions in the field. Our conclusions could have dramatic practical implications, in the form of changed approaches to analysis of conflict resolution, with better guidance to researchers and conflict resolution practitioners alike.

Keywords: International Conflict Resolution · Data mining · War and Peace

1 Introduction

Babbitt and Hampson (2011, p. 46) provide a description and critique of the field of International Conflict Resolution (ICR).

> "Theory and research," they argue, "are drawn not only from political science but also from social psychology, sociology, economics and law... IR [sic] scholars perceive a bias among Conflict Resolution scholars and practitioners towards peaceful methods of dispute settlement and resolution, one that deliberately and self-consciously eschews the use of force and violence."

The critique argues that there are inherent biases affecting conflict resolution practitioners and theoreticians in their approach to this field. As a result, they suggest that the field of International Conflict Resolution research may, by reasons of ideology, philosophy or background familiarity, be substantially affected by unscientific and inappropriate biases in research.

Babbitt and Hampson go on to posit that a more genuine analysis of International Conflict Resolution should be as two interrelated fields of study and endeavour- "conflict settlement" and "conflict transformation." Thus the goal of "International Conflict Resolution" in its entirety is to determine what processes and procedures are most necessary to achieve a maximisation of peace through conflict prevention, peacemaking through the most efficient and best methods of resolving current conflicts

© Springer Nature Switzerland AG 2019
D. C. Morais et al. (Eds.): GDN 2019, LNBIP 351, pp. 123–136, 2019.
https://doi.org/10.1007/978-3-030-21711-2_10

and the creation of stable political and legal structures so as to avoid the prospect of conflict, in the form of war or violence, arising in future.

However, the achievement of "peace" and "conflict resolution" are not the same thing. More importantly, "peacefully obtaining an outcome" and "resolving a conflict" are very different things. The former places conditions on a method of achieving a result. It rules out the prospect of war, armed conflict or, likely, the threat of such in order to obtain a political or practical settlement, i.e. an outcome whether formalised between the parties, or merely a detente.

If Conflict Resolution is to be seen as the combined fields of "conflict settlement" and "conflict transformation", as opposed to being akin to the much narrower "peace studies", (defined by Samaddara (2004) as merely "the study of peace and mechanisms to bring about peace as an active pursuit"), due consideration should be given to all methods actually or potentially utilised by states and other participants in international conflict resolution to achieve resolution of disputes. It would appear inappropriate to predetermine which methods are "legitimate" for the international community to use. This, however, is the norm within the academic disciplines of International Conflict Resolution, International Dispute Resolution, Peace Studies and other interrelated fields, as outlined below.

Addressing this issue is complicated by the ongoing blurring of terminology used in the consideration of conflicts, as reflected by a number of increasingly interrelated disciplines. Literature associated with the determination and management of international "quarrels" contains references to fields of studies and concepts including International Conflict Resolution, International Dispute Resolution, International Conflict Settlement and International Dispute Settlement, as well as "Peace Studies", "Peace Research", "Conflict Management" and others.

As is apparent, there are few "neutral" words that are not already embraced by the literature. The selection of "quarrel" is not to suggest a new term, but a generalized reflection of disputes, conflicts, clashes, arguments, etc. Terminologically, "conflict" and "dispute" are distinct ontological terms, the former indicating issues that are not negotiable and the latter indicating "negotiable interests." (Burton 1991, p. 62) "Settlement" and "resolution" are also distinct, with the former, classically, referring to negotiated outcomes, rather than "resolution", which Burton defines to mean "outcomes of a conflict situation that must satisfy the inherent needs of all." Were these definitional distinctions to be applied in practice, a number of distinct fields of study would exist within the matrix of conflict-dispute and settlement-dispute dichotomies alone, as outlined below (Table 1).

Table 1. Field Definitions

Field Definitions- International "Conflict Settlement", "Conflict Resolution", "Dispute Settlement" and " "Dispute Resolution"		
	Settlement	**Resolution**
Conflict	Negotiated outcomes to disputes that are not negotiable	Achievement of outcomes that are generally satisfactory to all parties over issues that are non-negotiable
Dispute	Negotiated Outcomes to negotiable interests	Achievement of outcomes that are generally satisfactory to all parties over issues that are negotiable.

Even a cursory analysis of the literature shows that these fields, if ever separate and distinct, have functionally merged through the misuse of terminology. "International Conflict Settlement" - the achievement of negotiated outcomes to disputes that are non-negotiable should not be possible and hence the term should have no meaning. If "conflicts' are successfully negotiated, they should by definition have been considered to have actually been "disputes". However "conflict settlement" is a popularly used term. Lieberfeld (1995, p. 201), Hannah (1968, p. 1) and Dixon (1994, p. 32) all illustrate the sustained use of "conflict settlement" and its seeming interchangeability with "conflict resolution", "dispute resolution" and "dispute settlement" as descriptors of issues between parties that may or may not be negotiated or negotiable. Hadzi-Vidanovic (2010) describes the role of the court as "conflict manager" and a resolver of international disputes. On this analysis, there is no functional differentiation between International Conflict Resolution ("ICR") and International Dispute Resolution ("IDR") and certainly no applied distinction between either of these terms and International Conflict Settlement ("ICS") or International Dispute Resolution ("IDR").

On any construction of the above definitions, war and armed conflict are very much on the outer, applicable only to an analysis, at most, of conflict. Even then, the role of war in "determining" or "ending" conflicts or disputes would seem to be excluded by definition. Rather, war would appear to be only a matter for consideration as a "conflict" in and of itself, rather than a method of "resolution".

Accordingly, it is unsurprising that war and armed conflict are, very little analysed in terms of their effectiveness in resolving conflicts within ICR. This is a deeply troubling outcome given the frequency with which war, or the threat of armed conflict, is actually used in international political negotiations and conflict resolution attempts. War and armed conflict have become increasingly central to international political affairs, with the Twentieth Century "the bloodiest epoch of all human civilization. The barbarism that characterizes the past hundred years is greater than any that afflicted earlier times" (Cheldelin et al. 2008, p. 9). ICR research does cover, at least to some extent, other tools that are available to states to enforce resolutions or to pressure states to behave in certain ways. One such tool is the application of sanctions. (Amley 1998, p. 235) However, the study and consideration of the use of military force, or the threat thereof, as tools in conflict resolution remains anathema across the field.

This paper aims to put the consideration of war into the proper context within International Conflict Resolution. Firstly, we survey the current extent of "coverage" offered by ICR, both of kinds of conflicts addressed and the methods of conflict resolution or transformation generally studied within the field. Secondly, we consider the role that military action and the threat of military action (generally termed "war" within this paper for convenience), play in international conflicts and conceptually in ICR. Thirdly, we consider the degree to which war or militarized action is, in fact, efficiently used in international affairs. Finally, we consider the degree to which ICR, in both its theoretical constructs and in statistical research, considers war.

2 The Extent of the Use of Negotiation, Mediation, Arbitration and Other Methodologies in International Conflict Resolution

Each of negotiation, mediation and arbitration, in a variety of permutations, has been widely used in conflict resolution (Lodder and Zeleznikow 2010). Bercovitch's study of modern conflict resolution in armed conflicts identifies more than 300 separate conflicts, many containing more than 20 different attempts to resolve the conflict (Bercovitch 2004). However, different approaches and data codification approaches can lead to different criteria for consideration. Relative determinations of the numbers of attempts at negotiation or mediation of conflicts are therefore somewhat difficult to achieve. More than 150 cases have been referred to the International Court of Justice since its inception in 1945 (International Court of Justice Advisory Opinions by Chronological Order 2014).

Since its establishment in 1997, the International Tribunal for the Law of the Sea has heard 22 cases (International Tribunal for the Law of Sea, Cases, 2014)[1]. The Permanent Court of Arbitration, established in 1899, has heard at least 40 state-state cases, as well as many cases involving states and non-state actors[2] (Permanent Court of Arbitration, Past Cases 2014). Determining the actual extent of the use of methods of conflict resolution is also a matter of extensive methodological dispute - defining what constitutes a mediation or a negotiation, separating out each attempt and otherwise identifying matters that are not necessarily in the public domain, can be extremely difficult.

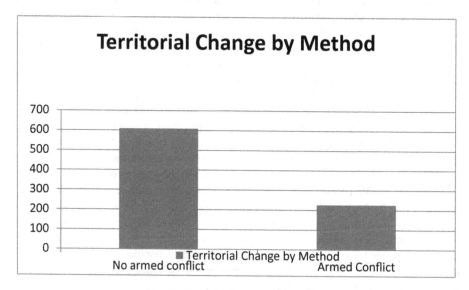

Fig. 1. Territorial change by method

---

[1] See https://www.itlos.org/fileadmin/itlos/documents/cases/case_no.21/verbatims/ITLOS_PV14_C21_2_Rev.1_E.pdf last accessed 12/07/2018.

[2] Permanent Court of Arbitration, Past Cases, https://pca-cpa.org/en/cases/ last accessed 10/01/2018.

Figure 1 illustrates the extent to which armed conflict, in some form, is involved directly in territorial change events. Included in the above data are 122 acts of conquest (Tir et al.1998, p. 89). From 1945 until 2008, 21 cases of conquest and 56 cases of conflict were recorded as the basis for territorial change, a rate of 19%, which is appreciably lower than the conflict rate since 1816 of 27%. As a result, it seems clear that conflict accounts for a major mechanism in changes to geopolitics across the globe, and over a sustained period of time.

Figure 2, drawn from the Correlates of War Militarised Incident Data (Ghosn et al. 2004) considers the various militarised and quasi-militarised occurrences in interstate conflicts. Altogether, 3317 incidents were recorded in the period 1993–2010. Whilst it is impossible to ascribe armed action as necessarily indicating an attempt to resolve a dispute, as opposed to escalating a dispute, responsive efforts or action taken for other purposes, the data records 730 threats of hostile action. Threats, coupled with actual use of force or the creation of "facts on the ground", accounted for more than 2,200 events. On any measure, this is a substantial number of actions effecting international relations and conflicts.

It therefore seems reasonable to suggest that ICR should concern itself significantly with the operations of armed ICR, as well as with the impact that the potential availability of force may have on the decision-making processes of participants in all forms of ICR, even whilst engaged in ostensibly peaceful resolution.

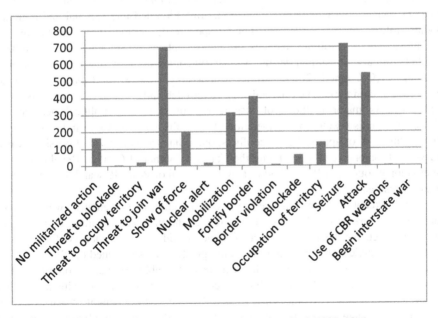

Fig. 2. Interstate conflict - mechanisms exercised, 1993–2010

3 How Is War Studied Within International Conflict Resolution?

Armed conflict plays a major role within ICR studies - as a substantial form of conflict which is sought to be resolved without violence. Armed conflict is viewed by many within the field as the worst form of conflict, occasioning a direct loss of human lives (Sohn 1983). As such, a particular emphasis is placed by many practitioners on resolving armed conflict, preventing the use of armed conflict and avoiding the escalation of disputes to the use of violence.

The successful resolution of a conflict is often ascribed as the cessation of armed hostilities, even if the underlying issues remain. As such, armed conflict, in a very real sense, underpins the foundations of many negotiations over matters of significant conflict. The capacity to initiate armed conflict is also, classically, one of the prime differentiators of states from other non-state international actors. Whilst the avoidance of armed conflict may not necessarily result in the overall preservation of human life, the measure of success in ICR is often the elimination or reduction in directly attributable deaths, regardless of other outcomes. This is highly questionable, both from the perspective of being a realistic measure of the impact of a conflict, and in terms of making progress towards the finalisation of a conflict (Stern and Druckman 2000, p. 4).

A review of works and journals on conflict resolution reveals little research into the role and impact of armed conflict on international conflict resolution. By way of illustration, principal texts, such as Merrills' *International Dispute Settlement* (2011) contains sections on negotiation, mediation, arbitration and international and regional bodies' role in dispute settlement, but no specific discussion of war or the use of military power in resolving disputes. Its index contains no reference to war, armed conflict, military, or similar terms. Similar outcomes exist in International Dispute Settlement (O'Connell 2003), which canvasses a range of tribunals, political manoeuvres and international frameworks. Bercovitch and Jackson (2009) also isolates and considers major methods of dispute resolution from a political perspective. However, there is next to no consideration of the realistic use of war or military force in the resolution of or approaches to conflicts, other than to avoid it in all possible ways. Surveys of major journals, including the Cardozo Journal of Conflict Resolution, The International Journal of Peace Studies and the Journal of Peace Research also reveal no relevant results.

A rare exception to this silence is the consideration of preventative war. Preventative war, or armed conflict designed to reduce the prospect of future, more violent armed conflict, has been the subject of some consideration. Armed conflict is considered as an undesirable outcome, with only greater intensity armed conflict (likely to result in *greater* loss of life) considered more undesirable. Similarly, some consideration has entered the field as to the broader desirability of avoiding or reducing conflict by strengthening defensive measures so as to both reduce the benefit of war as a tool, and so as to change parties' bargaining positions. In this sense, an oblique recognition of the role of armed conflict in negotiation does arise. However, these represent slight, almost tangential considerations in comparison to the centrality of armed conflict, and the capacity for armed conflict, in interstate relations.

The dearth of research into war within the field of International Conflict Resolution Studies is deeply surprising, given that there is a broad range of research into war, into the relationship between war and politics and into belligerent foreign policies. The tradition of studying war as a political tool extends back at least as far as Sun Tzu, Machiavelli and von Clausewitz. This tradition, however, does not meaningfully manifest itself in ICR. *The Handbook of War Studies*, offers a wide range of approaches to the analysis and statistical study of war (Midlarsky 1989). However, even in this book, war is considered in its own right, rather than as a tool for conflict resolution.

Singer (1989, pp. 12–14) explores the concept of war as a rational decision; as a product of realpolitik, in which states acting in their own interests may choose to employ armed conflict as a tool of policy. Doran (1989) explores the role of military capability in foreign policy as part of the cycle of the rise and fall of nations (at p. 85), but stops short of any consideration of how armed conflict itself is used in resolving conflicts. Others, such as Withana (2008), offer a broader analysis of war, considering the no law approach and the consequent use of force by states. She cites a line of political and military thinkers who consider "the application of force as the "… most effective contribution towards achieving the ends set by political policy…" She offers further consideration of some of the limitations and impacts on the possible state behaviours and methods of conflict resolution occasioned by military capabilities. However, even this consideration avoids consideration of any actual role for armed conflict in dispute resolution. We have found no substantive consideration of the role played by armed conflict or the threat of armed conflict in international conflict resolution. Other than the occasional tangential mention or reference linking the fields, a broad gulf exists between studies and the understanding of armed conflict and its integration into our understanding of international conflict resolution.

4   A New Framework for the Effectiveness of International Conflict Resolution

Having reviewed the existing challenges and limitations of the consideration of armed conflict within ICR, we propose the adoption of a new framework for the analysis of the effectiveness of ICR attempts and methods. We hypothesize that the principal challenge facing ICR, and the primary resistance to the proper analysis of force within ICR activities, is the struggle to meaningfully answer the following question:

How should we evaluate the relative worth of different outcomes of a conflict?

To answer the questions requires an evaluation of what are basically the preferences between a number of potential values, which can themselves be broadly characterised as efficiency values, or subjective values. A partial list of factors might include:

Efficiency Values	Subjective Values
(1) Finality of the Resolution (2) Speed of the Resolution (3) Economic Cost of the Resolution (4) Political cost of the Resolution (5) Social Cost of the Resolution (6) Probability of Recidivism Following Resolution Prevention of escalation of the conflict to higher level	(1) The compliance of the resolution with international law (2) The '"justice" of the outcome (3) The adherence of the outcome to social or cultural norms (4) Asymmetry in costs and benefits to the parties The methods used in achieving the outcome

Even the Efficiency Values, which are potentially empirically measurable, are inherently matters of preference. For instance, for business, the relative interest in achieving finality may outweigh speed, or at least be of different importance to different groups. The prospect of "peace in our time" may be attractive to some, whilst anathema to others. Like an elderly neighbour not wishing to spend lavishly to fix a building's foundations and prevent a distant future collapse, the way that potential resolutions to disputes can be evaluated requires us to make, and identify that we are making, a choice.

The presentation and classification of these factors is, to our knowledge, novel, though there are some suggestions within the International Correlates of War data that suggest that others have considered research along these lines. For instance, the ICOW territorial claims codes for a number of measures that could be used to determine finality, such as whether the dispute reoccurs within three years, speed of the resolution and the highest level of the dispute.

Thus far, ICR studies has generally made a fundamental assumption that one subjective factor, the method used in achieving the resolution, should carry prime importance. This is despite a lack of application of empirical evidence that shows that outcomes that are achieved without war are necessarily better, at least insofar as the effectiveness of the outcome is concerned. This is also despite the obvious reality that many countries are willing to use force to achieve their goals, having determined that the subjective value of the avoidance of war is overweighed by other interests.

We therefore suggest that the field of ICR should have a renewed emphasis on consideration of which methods are most favourable to achieving each kind of outcome, rather than on the presupposition of a particular evaluation of the relative worth of each of the values. It may be possible to develop scales that weigh up a number of efficiency values as well, though this in itself becomes a value judgement. We suggest, though, that as a first stage, this will broaden the thinking of ICR and encourage a better, more rigorous approach to considering how parties behave in ICR attempts, based on a true evaluation of parties' individual interests.

ICR researchers already make use of a broad range of tools of study in their research into each method of conflict resolution, and in the consideration of the vast range of hybrid methodologies and systems that exist within these broad categories. Principally, case-based analysis of both international dispute resolution attempts and more broadly international negotiations has assumed a great deal of popularity (Druckman 2005, p. 163). Case-based analysis is popular because of the uniqueness of each international conflict, as well as the level of detail that is facilitated by this approach.

International Conflict Resolution researchers now make use of statistical research into conflicts. Statistics has assumed a lesser role than other techniques, because of a number of factors. Firstly, given the long history of international disputes, determining or implementing a single set of standards or range for inclusion within datasets is itself potentially a matter of huge debate. Secondly, given the complexity of international disputes and conflicts, there are significant concerns as to the relevance of summarily codified data. Thirdly, philosophically, the statistical study of conflicts has not always accorded well with many researchers, as outlined below.

However, in recent years, projects such as the International Correlates of War dataset on armed conflicts, territorial conflicts and maritime boundary disputes, and Bercovitch's work on mediation and dispute resolution attempts, have become available to researchers. These build on a solid but sparse history of attempts to categorically record and codify sets of disputes, going back to the 1960's, and to analyse their outcomes, with consequent potential for research Singer (1989, p. 2).

There has been little to no study focussed on selecting the most suitable conflict resolution method for particular conflicts (Bercovitch and Jackson 2001, p. 59). For example, territorial disputes between middle powers could be resolved in many cases either by reference to international arbitral bodies, presupposing a treaty-based mechanism for such a referral, by mediation, direct negotiation or through third-party facilitation. Optimisation of conflict resolution, as well as each party's interests, may be best served by adopting both particular conflict resolution strategies (or non-resolution strategies) as well as by agreeing to refer the matter to resolution through a particular processes. A genuine and full understanding both of the factors within each method and globally, which lead to success or failure, is the naturally desirable outcome in order to achieve both the goals of conflict settlement and conflict transformation. This can only truly happen, though, when the evaluation of the preferability of the outcome is separated from the evaluation of the efficiencies of the outcome i.e. for what conflicts is this method suitable?

However, such developments represent a long-term vision for the field of ICR, and are not an immediate prospect, given the state of the field. Research undertaken by Bercovitch and Jackson (2009) indicates that states are likely to use methods based on their availability, contextual factors, the complexity of the issue and the desirability to parties of reaching a resolution altogether. As a result, parties rarely make a decision on the basis of the most effective method in actually resolving the conflict, instead adopting an ad-hoc approach to selecting a method to use. If there was real research into the efficacy of different methods in achieving outcomes optimised to each efficiency value, this might be radically changed.

4.1   Illustration One: The Question of Compliance

An illustration of the relative valuation of different outcomes in ICR is the issue of compliance - the enforcement, and enforceability of agreements reached through ICR interventions. The entrance by parties into formal terms of resolution of a dispute does little to change facts on the ground. The impact of an agreed resolution, or a nominally binding ruling, on the actual power positions of the respective parties is largely legal and political. Whilst some reputational damage is likely to result when parties refuse to comply with their obligations, unless the ruling can be enforced, it is unlikely to have any direct impact. International enforcement of awards is notoriously unpredictable. The International Court of Justice, for instance, is dependent upon the United Nations Security Council for the capacity to enforce any determination (United Nations Charter Chapter XIV). However, ultimately, enforcement of international agreements requires the means to do so. This is often a political question, rather than a process question, and is therefore regularly avoided in the analysis of international dispute resolution, as being beyond the scope of analysis.

For many ICR researchers and practitioners, the measure of success is the entrance into an agreement, rather than subsequent adherence to the agreement. This is obviously a different valuation to that which other parties, such as business interests, may adopt, for whom "facts on the ground" carry far more weight than formal international instruments.

4.2   Illustration Two: Enforcement of Arbitration Outcomes

A range of methods of enforcement are available and are utilised by states in international disputes. Many methods are the product of international treaties, such as punitive sanctions emergent from General Agreement on Tariffs and Trade (GATT) and World Trade Organisation (WTO) processes.[3] Other methods, such as a declaration by a relevant court or tribunal, carry both legal weight and political consequences for countries involved. Broadly speaking, though, such responses can be described as framework responses and enforcement approaches - methods of seeking redress that require either some degree of compliance by the wrongdoer, or continued adherence with international law, norms and agreements. These responses, as well as other treaty-based methods of enforcement or creation of international law such as injunctions, referral to the International Court of Justice or even soft power or international political disapproval, are next to useless if a country is consistently non-compliant, is insulated or otherwise indifferent to the international protocols involved.

Pariah-states such as North Korea are arguably indifferent to international attitudes, whilst other states, such as Russia, may prioritise particular national interests over the impact of sanctions that might be applied. Unlike a normative judicial enforcement process in municipal law, a country or countries seeking to enforce a decision must be

---

[3] *Understanding on Rules and Procedures Governing the Settlement of Disputes,* World Trade Organisation, see https://www.wto.org/english/tratop_e/dispu_e/disp_settlement_cbt_e/c1s2p1_e. htm last accessed 10/03/2018.

willing to face consequences themselves, such as reduced economic benefit, where sanctions are used.

Stricter tools available to countries, in some circumstances, include economic sanctions, embargoes, political sanctions and the application of soft-power. Depending on the relative political, economic and geopolitical positions of the parties, some of these methods can be highly effective.

For instance, the use of concerted embargoes on the sale of oil in the 1970's was highly effective at influencing Western nations to change their policies towards Israel. On the other hand, the capacity or effectiveness with which states can attempt such measures is connected with their own strength and political support. Sanctions often require cooperation and in today's global economy, are rarely effective without a broad base or backing. Recent US-led sanctions on Iran are a powerful illustration of the necessity to build broad coalitions amongst both states and political institutions so as to effectively apply enforcement mechanisms or sanctions against other states.

Many states are also capable of taking direct action against other states, either as a method of enforcement or as a method of establishing facts on the ground as a way of resolving or transforming international conflicts. Armed conflict or war is amongst the most basic methods of achieving this goal. For example, territorial disputes can be resolved or altered by one state unilaterally claiming and garrisoning the disputed territory. The establishment of a strong military position and the demonstrated willingness to carry out further military actions has been a staple of foreign policy throughout the modern era. The *International Correlates of War Militarized Interstate Disputes* Database (Ghosn et al. 2004) records more than 1700 instances of military action or formal threats of military action in the decade following the creation of United Nations. In the period from 1816 to 1945, less than 800 such threats or actions are recorded in the database, as having occurred. As such, armed conflict is inescapably part of the modern international dispute resolution process, with hundreds of cases occurring.

War is also used as a way of compelling countries to adopt internal policies in adherence with those of other states. The NATO intervention in the Kosovo conflict involved the enforcement of international norms and Security Council resolutions through a campaign of applied state-based violence against the Serbian military and economy, including the bombing of key facilities and even cultural and civilian targets Amley (1998, p. 240). Conflict between the Russian Federation and its near-neighbours in recent years has seen Russian troops used to resolve policy conflicts over the independence aspirations of Russian speaking regions in both the former Soviet republics Ukraine and Georgia.

Nations such as Syria and Iran have long-funded proxy-groups in attacks on Israel, in an attempt to weaken the latter state, gain political hegemony in the region and to bolster political and military positions in territorial negotiations. War and armed conflict can therefore be reasonably described as using widely available tools through which states advance their policy goals, resolve conflicts, transform the on-the-ground realities of conflicts, enforce compliance with agreements and even as a threat to provoke concessions in negotiations between parties.

The threat of war, too, plays a major role in conflict resolution. As noted, escalating numbers of threats of armed conflict have been made in communications between states since the creation of the United Nations. These, distinct from actual acts of violence, indicate a willingness on the part of states to employ armed conflict on the one hand, but also the potential for armed conflict as a stated alternative to a resolution. The threat of armed conflict need not be explicit for it to influence negotiations. As such, the influence of the potential for armed conflict on negotiations is potentially much larger than the number of specifically noted instances indicates. During negotiations, states can also make use of the potential for armed conflict in many ways. For example, the use of military aid to states underpins much of American foreign policy and influence. It also buys American influence and a seat at negotiating tables in a range of conflicts. The commitment of military assistance in the event of a future conflict also underpins a range of treaties on non-military matters, and helps ensure that political and economic relationships are stabilised and secured. Equally, the promise to refrain from making use of armed conflict, influences relationships and is a major tool in actually resolving conflicts. Thus, both the use of armed conflict, and the absence of armed conflict, influence negotiations and a wide range of efforts at conflict resolution.

It is also apparent that the enforceability of an arbitration outcome is of differing importance to the various parties. For Australia, its ability to enforce its victory over Japan in the Whaling Decision[4] or to seek damages from Japan, is of far lesser significance than the moral and political pressure that Australia is now able to exert. By contrast, the principal interests of South American countries in arbitrating their border disputes may well have been the achievement of enforceable and legally binding outcomes, which may in turn have had economic and/or social consequences within their borders.

5 Conclusion

International Conflict Resolution Studies incorporates a broad range of sub-fields and draws from the expertise of a number of disciplines. The actual operation of International Conflict Resolution studies has focused almost exclusively on political and legal methods of conflict resolution, primarily negotiation, mediation and arbitration. However, amongst the most prevalent methods of conflict resolution actually employed by states, and underpinning many negotiated or mediated International Conflict Resolution attempts, is armed conflict. International Conflict Resolution studies has operated almost to the complete exclusion of the consideration of the effectiveness or operation of armed conflict as a method of ICR. This has resulted in the development of a limited understanding of both state motivations and negotiating strategies, as well as developing a major lacuna in the understanding of international conflicts.

Ultimately, the goal of ICR must be to be able to fully optimise conflict resolution by applying appropriate methodologies to appropriate conflicts. The goal of ICR as a field of study must accordingly be to develop the knowledge and understanding to enable practitioners to apply the best method and at the best time, depending on the

---

[4] International Court of Justice Decision, Whaling in the Antarctic (Australia V. Japan: New Zealand Intervening), Judgement 31 March 2014, viewed 1 April 2015 via http://www.icj-cij.org/docket/files/148/18136.pdf last accessed 12/07/2018.

relative valuation of outcomes and the probability of success. Whilst there is much work to be performed to develop our understanding of each methodology, any serious comparative analysis must include the full range of options which states may utilise if a peaceful conflict management attempt is unsuccessful. To do otherwise is, ultimately, likely to lead to inaccurate conclusions.

International Conflict Resolution studies can be both aspirational, in the sense that it seeks to enable states to resolve conflicts efficiently and non-militarily, and rigorous, in the sense that it can consider the actual operative nature of international conflict resolution. Babbit and Hampson (2011) also note that:

> Policy makers and CR practitioners still have a tendency to overlook or dismiss research findings. In part, this is because of the inevitable differences in professional cultures between academia and the so-called "real world."

So long as sharp gaps remain in what International Conflict Resolution researchers are willing to consider insofar as the actual realities that face practitioners, the usefulness of International Conflict Resolution research is likely to be severely limited.

References

Amley, E.: Peace by other means: using rewards in UN efforts to end conflicts. Denver J. Int. Law Policy **26**, 235 (1998)

Babbit, E., Hampson, F.: Conflict resolution as a field of inquiry: practice informing theory. Int. Stud. Rev. **13**, 46–57 (2011)

Bercovitch, J.: International Mediation and Intractable Conflict (2004)

Bercovitch, J., Jackson, R.D.W.: Conflict Resolution in the Twenty-First Century: Principles, Methods, and Approaches. University of Michigan Press, Ann Arbor (2009)

Burton, J.: Conflict resolution as a political philosophy. Glob. Change Peace Secur. **3**(1), 62–72 (1991)

Cheldelin, S., Druckman, D., Fast, L., Clements, K.P.: Theory research and practice. In: Cheldelin, S., Druckman, D., Fast, L. (eds.) Conflict: from Analysis to Intervention, 2nd edn, pp. 9–35. Continuum, New York (2008)

Dixon, W.J.: Democracy and the peaceful settlement of international conflict. Am. Polit. Sci. Rev. **88**(01), 14–32 (1994)

Doran, C.F.: Power cycle theory of systems structure and stability: commonalities and complementarities. In: Handbook of War Studies, pp. 83–110 (1989)

Druckman, D.: Doing Research: Methods of Inquiry for Conflict Analysis. Sage, Thousand Oaks (2005)

Ghosn, F., Palmer, G., Bremer, S.A.: The MID3 data set, 1993–2001: procedures, coding rules, and description. Conflict Manage. Peace Sci. **21**(2), 133–154 (2004)

Hadzi-Vidanovic, V.: Conflict settlement by the international court of justice (2010). https://papers.ssrn.com/sol3/papers.cfm?abstract_id=1988922. Accessed 7 Dec 2018

Hannah, H.: Some dimensions of international conflict settlement procedures and outcomes (No. RR-11). International Court of Justice, The History of the Court, Department Of Political Science, Hawaii University Honolulu (1968). http://www.dtic.mil/docs/citations/AD0727152. Accessed 3 Oct 2018

Lieberfeld, D.: Small is credible: Norway's niche in international dispute settlement. Negot. J. **11**(3), 201–207 (1995)

Lodder, A., Zeleznikow, J.: Enhanced Dispute Resolution Through the Use of Information Technology. Cambridge University Press, Cambridge (2010)

Midlarsky, M. (ed.): Handbook of War Studies. Unwin Hyman, Boston (1989)

Merrills, J.G.: International Dispute Settlement. Cambridge University Press, Cambridge (2011)

O'Connell, M. (ed.): International Dispute Settlement. Dartmouth Publishing Company, Aldershot (2003)

Samāddāra, R.: Peace Studies: An Introduction to the Concept, Scope, and Themes [e-book]. Sage Publications, New Delhi (2004)

Singer, J.D.: System structure, decision processes, and the incidence of international war. In: Midlarsky, M.I. (ed.) Handbook of War Studies, pp. 1–21 (1989)

Sohn, L.: The future of dispute resolution. In: Macdonald, R.S.J., Johnston, D.M. (eds.) The Structure and Process of International Law: Essays in Legal Philosophy, Doctrine and Theory, vol. 6, pp. 1121–1146. Martinus Nijhoff Publishers, Leiden (1983)

Stern, P.C., Druckman, D.: Evaluating interventions in history: the case of international conflict resolution. Int. Stud. Rev. 2(1), 33–63 (2000)

Tir, J., Schafer, P., Diehl, P.F., Goertz, G.: Territorial changes, 1816–1996: procedures and data. Conflict Manage Peace Sci. 16(1), 89–97 (1998)

Withana, R.: Power, Politics, Law: International Law and State Behaviour During International Crises. Martinus Nijhoff Publishers, Boston (2008)

The Effect of Conformists' Behavior on Cooperation in the Spatial Public Goods Game

Yinhai Fang[1,3] , Haiyan Xu[1(✉)] , Matjaž Perc[2,3(✉)] ,
and Shuding Chen[1]

[1] College of Economics and Management, Nanjing University of Aeronautics
and Astronautics, 169 Sheng Tai West Road, Nanjing 211106, China
xuhaiyan@nuaa.edu.cn
[2] Faculty of Natural Sciences and Mathematics, University of Maribor,
Koroška cesta 160, 2000 Maribor, Slovenia
matjaz.perc@gmail.com
[3] CAMTP-Center for Applied Mathematics and Theoretical Physics, University
of Maribor, Krekova 2, 2000 Maribor, Slovenia

Abstract. In this paper, we investigate the effects of rational and irrational conformity behavior on the evolution of cooperation in public goods game. In general, conformist should also probably consider the difference of payoff between himself and his neighbors. Therefore, we divide the players into two categories: traditional payoff-driven players and secondly, rational conformists. Rational conformists will only update their strategy according to the conformity-driven rule when they get a higher payoff than their neighbors, whereas irrational conformists' updating rule is the opposite. Remarkably, we find that both rational and irrational conformists enhance cooperation in the spatial public goods game. However, the differences in intensity of this positive effect between rational and irrational conformists are tremendous, and the latter promotes a higher level of cooperation to reach a much higher level and extensive positive effect.

Keywords: Public goods game · Conformity behavior · Social dilemmas · Cooperation

1 Introduction

Cooperation and defection are two key strategies usually existing at the heart of every social dilemma [1–4]. In the areas of environmental resources or social benefits, defectors usually reap benefits on the expense of cooperators. The "tragedy of the commons" succinctly describes such a situation [5, 6]. In the last two decades, evolutionary game theory [7, 8] has strongly developed into a powerful tool for modelling a myriad of social dilemma phenomena characterized by evolutionary dynamics and complex interaction patterns [9, 10]. And ample researchers have focused on the identification of mechanisms that may lead to high cooperation.

Payoff maximization as a classic mechanism has been extensive studied in the past [11–13]. The key assumption behind these researches have been that each player only

© Springer Nature Switzerland AG 2019
D. C. Morais et al. (Eds.): GDN 2019, LNBIP 351, pp. 137–145, 2019.
https://doi.org/10.1007/978-3-030-21711-2_11

aspires to maximizing its own payoff. It is not always the case in many real-life situations. An individual tends to be follow the majority in behavior or opinion within the interaction range and the conformity also plays an important role in the society [14]. Perc and Szolnoki designated a fraction of population as conformity-driven players instead of payoff-driven players [15, 16]. The conformity-driven players adopt their strategies simply according to the popularity of strategies among populations in their research. They showed that an appropriate fraction of conformists will introduce an effective surface tension around cooperative clusters and ensures smooth interfaces between different strategy domains. Yang and Tian proposed a conformity-driven reproductive ability in which the probability that an individual adopts a strategy both depends on the payoff difference and the popularity of strategies [17]. They find that the cooperation level can be enhanced by moderately increasing the teaching ability of the neighbor with the majority strategy in the local community. Javarone and Antonioni [18] studied the spatial public goods game in the presence of social influences considering both conformity-driven players and fitness-driven players. They find that conformism drives the system towards ordered states, with a prevalence for cooperative equilibria. Niu and Xu [19] set the rational conformity behavior by introducing the mechanism that player will compare its payoff to his last time step payoff. If its payoff at this time is worse than last time step then it will tend to adopt the most common strategy of its neighbors. Yang and Huang [20] treat strategy-updating rule (payoff-driven or conformity-driven) as an attribute of players and allow for the evolution of the attribute and find that frequent alternations of the strategy-updating rule with unbiased rule enhances cooperation.

Motivated by the previous work, we consider two different behaviors of conformists, in which the individual i adopts a randomly chosen neighbor j's strategy with the probability driven by conformity only after comparing their payoff at first. Not the same as researches in [17] set all the players in the population has only one rule. Here, we consider the payoff-driven rule, rational and irrational conformity-driven rules. It is also worth noting that, the rational and irrational behaviors defined in this work is by comparing the payoff of neighbors instead of their own payoff [19].

2 Model

We study evolutionary PGGs in a population of N players distributed uniformly at random on a square lattice with periodic boundary conditions. Each individual on site x is designated either as a cooperator ($s_x = C$) or defector ($s_x = D$) with equal probability. They play the game with their $k = 4$ neighbors and each of them belongs to $G = 5$ different communities. It means that an individual is the focal individual of a Moore neighborhood and a member of the Moore neighborhood of its four nearest neighbors.

In a pairwise interaction, the cooperator contributes 1 to the public good while defectors contribute nothing as the standard parametrization. The sum of all contributions is multiplied by the synergetic factor $R > 1$, and the resulting amount is shared among the $k + 1$ interacting individuals equally regardless of their strategies. Denoting the number of cooperators and defectors among the k interaction partners by N_c and N_d respectively, each cooperator or defector gets payoff in one group as follows

$$P_c = R(N_c + 1)/(k + 1) - 1, \tag{1}$$

$$P_d = R(N_c + 1)/(k + 1). \tag{2}$$

Obviously, total payoff π_i of each player i is the sum of payoff got in five different communities it belongs to.

We consider two types of strategy-updating rules. One is the traditional payoff-driven rule and the other is the conformity-driven rule. The payoff-driven player i adopt to the strategy of their randomly chosen neighbor j with the probability determined by Fermi function [21]

$$W_{s_i \leftarrow s_j} = 1/\{1 + \exp[(\pi_i - \pi_j)/K]\}, \tag{3}$$

Where $\pi_i(\pi_j)$ is the payoff of player $i(j)$ and K quantifies the intensity of the noise related to the strategy adaption. In the conformity-driven rule, we consider two different behaviors of conformists. One is the rational conformist. The player i will compare its payoff with the randomly chosen neighbor j. Nothing will happen and player i will stick to its original strategy if $\pi_i \geq \pi_j$. On the contrary, player i will adopt the strategy of player j according to the probability of conformity-driven rule. The second behavior is the opposite of rational conformist and we set it as irrational conformist. The irrational conformist will adopt the most common strategy among its neighbors only when its payoff is better than its neighbor. It is worth pointing out that irrational conformist doesn't mean that the player is a stupid individual and it is just one kind of behavior. In real-life situations, irrational conformist may get more payoff than rational individuals at sometimes. The conformity-driven probability is described as [15]

$$W_{N_{s_i} - k_h} = 1/\{1 + \exp[(N_{s_i} - k_h)/K]\}, \tag{4}$$

where N_{s_i} is the number of players holding the strategy s_i in the neighbors of player i and k_h is one half of the degree of player i.

We simulate the model in accordance with the standard Monte Carlo simulation procedure. In this work, we set $N = 4 \times 10^4$, $K = 0.5$. Initially, the cooperators and defectors are randomly distributed among the population with equal probability. Payoff-driven rule and conformity-driven rule are assigned to players with probabilities of β and $1 - \beta$ respectively. We note that each full Monte Carlo step (MCS) consists of N elementary steps described below, which are repeated consecutively, thus making sure each player has the opportunity to change its strategy once on average.

(1) Select one player x randomly, and select one of its neighbors y randomly;
(2) Each player $x(y)$ plays the PGG with all its five different communities and then calculate the total payoff π_x and π_y;
(3) Player x performs the strategy revision phase according to its attribute, i.e. payoff-driven or conformity-driven (rational or irrational behavior).
(4) Repeat from (2) until N Monte Carlo steps elapsed.

3 Results

First, we study the impact of the number of rational conformists among the population on cooperation in the public goods game with different synergetic factor (R). In Fig. 1, we plot the evolution of the fraction of cooperators (ρ) as a function of MCS time for several different values of β and R. One can see that, for small values of R (*i.e.*, $R = 2, 3, 4$), the introduction of rational conformist has a little positive impact on the cooperation during the evolution process and the intensity of the effect increases with R increases. Especially, the system gets rid of all-Ds state with high fraction of rational conformist (*i.e.*, $\beta = 0.8$ and 1) when R increase to 4. It is worth noting that,

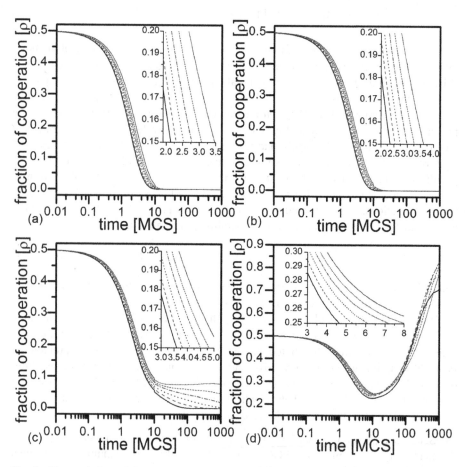

Fig. 1. The evolution of the cooperative frequency (ρ) with Monte Carlo time guided by the behavior of different fraction of rational conformists (β) among the population. (a)–(d) depict different synergetic effects of cooperation R = 2, 3, 4, 5 respectively. $\beta = 0$ (black solid line), $\beta = 0.2$ (red dashed line), $\beta = 0.4$ (blue dotted line), $\beta = 0.6$ (olive dash-dotted line), $\beta = 0.8$ (magenta short-dashed) and $\beta = 1$ (purple short-dotted line) are all considered with different combinations of $k_h = 2, N = 4 \times 10^4, K = 0.5$. To improve accuracy, the final results are averaged over 20 independent realizations, including the generation of random initial strategy distributions and rational conformist distributions, for each set of parameter values. (Color figure online)

this promotion loses its robustness and the system tends to be a high fraction of cooperation after experiencing a valley value (*i.e.*, $MCS = 10$, ρ is smaller than 20%) with the main effect of R as it increases to 5.

Then we depict the fraction of cooperators with the effect of irrational conformist within 1000 MCS in Fig. 2 in order to further explore what kind of conformist can boost the cooperation best. One can see that, the introduction of irrational conformist has a significant role in promoting cooperation when R is not so big (*i.e.*, R = 2, 3, 4). Figure 2(a), (b) show that ρ increases to 1 as β increase to 1 and the strength of positive effect is proportional to the fraction of irrational conformist among the population. We can also see that, for relatively small values of R (*i.e.*, $R = 2$ and 3), the positive effect

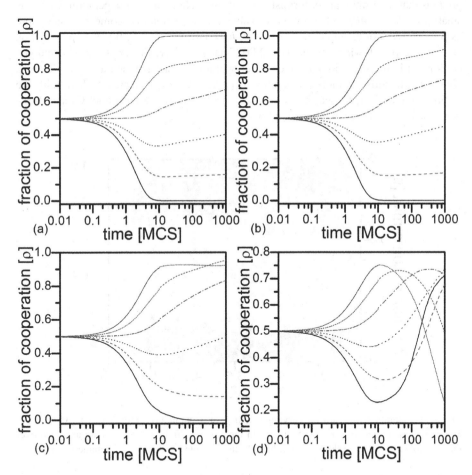

Fig. 2. The evolution of the cooperator frequency (ρ) with MCS considering the effect of different fraction of irrational conformists (β) among the population. (a)–(d) depict different synergetic effects of cooperation $R = 2, 3, 4, 5$ respectively. $\beta = 0$ (black solid line), $\beta = 0.2$ (red dashed line), $\beta = 0.4$ (blue dotted line), $\beta = 0.6$ (olive dash-dotted line), $\beta = 0.8$ (magenta short-dashed) and $\beta = 1$ (purple short-dotted line) are all considered with different combinations of $k_h = 2$, $N = 4 \times 10^4$, $K = 0.5$. Final results are averaged over 20 independent realizations, including the generation of random initial strategy distributions and rational conformist distributions, for each set of parameter values. (Colour figure online)

of β is robust as the main key of enhancement of cooperation no matter how R changes. However, for larger values of R (*i.e.*, $R = 4$ and 5), the positive role of β received the suppression of R. The highest fraction of cooperation $\rho = 0.95$ and 0.75 when $R = 4$ and 5 respectively. Especially, the effect of irrational conformist on cooperation is quite unstable and usually has a negative effect during the most period of evolution when $R = 5$. For example, the fraction of cooperation is $\rho = 0.23$ and 0.49 at the end of 1000 MCS when $\beta = 1$ and 0.8 respectively.

We depict the density of cooperator on varying the density of conformist (β) and the synergy factor (R) in Fig. 3 in order to complete a clear understanding about the effect of the fraction of irrational conformist on cooperative evolution. In general, it is obvious that the irrational conformist has a positive effect on the cooperation when R is small (*i.e.*, $R \leq 4.40$). When we fix the value of R, cooperators become more and more as β increases. For example, $\beta = 0$, $\rho = 0$; $\beta = 0.4$, $\rho = 47.9\%$; $\beta = 0.8$, $\rho = 94.5\%$, and $\beta = 1$, $\rho = 1$ when we fix $R = 2.80$. That is to say, the existence of irrational conformist in the population enables the cooperators to survive, and a large value of β could significantly promote cooperative behavior. When R is greater than 4.40, the positive impact is disturbed. Especially, $\beta = 1$ will no longer ensure all-Cs and disordered phase occurs when R is around 5.

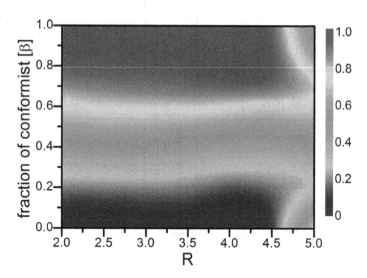

Fig. 3. Cooperation diagram on varying R and β in a population with $N = 4 \times 10^4$, $k_h = 2$, $K = 0.5$, R is in the range $\in [2.0, 5.0]$, β is in the range $\in [0.0, 1.0]$. Results are averaged over the last 3000 steps of 13000 MCS and have been computed using 21×5 parameter values.

To intuitively understand why the conformist that can affect cooperation when R is not so large (*i.e.*, $R = 2, 3$ and 4), we plot spatial strategy distributions as time evolves for traditional payoff-driven rule, rational conformity-driven rule and irrational conformity-driven rule from the top to the bottom respectively when $R = 2$. From Fig. 4(a1)–(a5), one can see that for the traditional payoff-driven rule, the defector cluster continually expands while the cooperator cluster rapidly shrinks. Similar pattern we can see in Fig. 4(b1)–(b5) but the speed of demise will be slightly slowed with the introduction of rational conformist (*i.e.*, MCS = 15 and 30 for the all-Ds state for traditional payoff-driven rule and rational conformity-driven rule respectively). However, the scenario is quite different in Fig. 4(c1)–(c5). Plenty of small cooperator clusters have been preserved and scattered throughout the whole population, which is very important for inhibiting the formation of defector cluster. In the end, the entire system tends to be an all-Cs state.

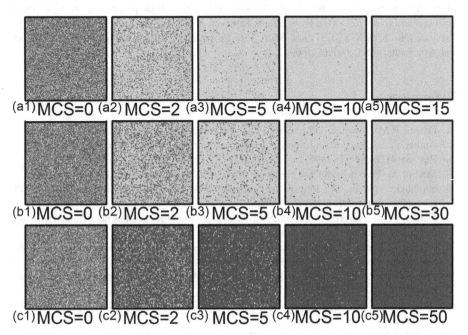

(a1) MCS=0 (a2) MCS=2 (a3) MCS=5 (a4) MCS=10 (a5) MCS=15

(b1) MCS=0 (b2) MCS=2 (b3) MCS=5 (b4) MCS=10 (b5) MCS=30

(c1) MCS=0 (c2) MCS=2 (c3) MCS=5 (c4) MCS=10 (c5) MCS=50

Fig. 4. Characteristic snapshots of cooperation (red) and defection (yellow) strategies with the effect of traditional payoff-driven rule, rational conformity-driven rule and irrational conformity-driven rule from the top to the bottom. The low synergetic effect of cooperation $R = 2$, as well as all results are obtained for $k_h = 2$, $N = 4 \times 10^4$, $K = 0.5$. (Colour figure online)

4 Conclusions

In evolutionary game theory, players usually update their strategies according to different strategy-updating rules. The traditional payoff-driven rule as one of the most popular rules has been widely studied in the last decades. Recently, more and more researchers pay attention to the conformity-driven rule. In the previous work [10], proper fraction of players following conformity-driven strategy-updating rule may improve network reciprocity and enhance cooperation. In this work, to further explore the impact of the conformist on cooperation, we have considered two kinds of behavior of conformists. In particular, the rational conformists are those that to imitate the strategy of the majority only when they got less payoff than their randomly chosen neighbor, while the irrational conformists are those that tend to imitate the strategy of the majority when their payoff are higher than their randomly chosen neighbor's. Here, we highlight the prominent role of irrational conformist in the spatial public goods game: it seems that rational and irrational conformist both enhance the cooperation among spatial public goods game. Whereas the latter promotes the population to reach a high level of cooperation and the positive effect is robust among a wide range of R.

Acknowledgments. This work was supported by the National Natural Science Foundation of China (Grants Nos. 71471087, 71071076 and 61673209), Major program of Jiangsu Social Science Fund (No. 16ZD008), Postgraduate Research & Practice Innovation Program of Jiangsu Province (No. KYCX18_0237) and Short Visit Program of Nanjing University of Aeronautics and Astronautics (No. 180908DF09).

References

1. Dawes, R.M., Messick, D.M.: Social dilemmas. Int. J. Psychol. **35**(2), 111–116 (1980)
2. Andreoni, J.: Cooperation in public-goods experiments: kindness or confusion? Am. Econ. Rev. **85**(4), 891–904 (1995)
3. Martin, A.: Five rules for the evolution of cooperation. Science **314**, 1560–1563 (2006)
4. Szolnoki, A., Perc, M.: Reward and cooperation in the spatial public goods game. EPL (Europhys. Lett.) **92**(3), 38003 (2010)
5. Hardin, G.: The tragedy of the commons. The population problem has no technical solution; it requires a fundamental extension in morality. Science **162**(3859), 1243–1248 (1968)
6. Feeny, D., Berkes, F., Mccay, B.J., et al.: The tragedy of the commons: twenty-two years later. Hum. Ecol. **18**(1), 1–19 (1990)
7. Perc, M., Szolnoki, A., Mccay, B.J., et al.: Coevolutionary games—a mini review. Hum. Ecol. **99**(2), 109–125 (1990)
8. Hofbauer, J., Sigmund, K.: Evolutionary game dynamics. Bull. Am. Math. Soc. **40**(4), 479–519 (2003)
9. Li, X.L., Jusup, M., et al.: Punishment diminishes the benefits of network reciprocity in social dilemma experiments. Proc. Natl. Acad. Sci. **115**(1), 30–35 (2018)
10. Axelrod, R.M.: The Evolution of Cooperation. Basic Books, New York (1985)
11. Griffin, A.S., West, S.A., Buckling, A.: Cooperation and competition in pathogenic bacteria. Nature **430**(7003), 1024–1027 (2004)
12. Biernaskie, J.M.: Evidence for competition and cooperation among climbing plants. Proc. Biol. Sci. **278**(1714), 1989–1996 (2011)
13. Turner, P.E., Chao, L.: Prisoner's dilemma in an RNA virus. Nature **398**(6726), 441 (1999)
14. Fiske, S.T.: Social Beings: Core Motives in Social Psychology, 3rd edn. Wiley, Hoboken (2014)
15. Szolnoki, A., Perc, M., Szolnoki, A., et al.: Conformity enhances network reciprocity in evolutionary social dilemmas. J. R. Soc. Interface **12**(103), 20141299 (2015)
16. Szolnoki, A., Perc, M.: Leaders should not be conformists in evolutionary social dilemmas. Sci. Rep. **6**(1), 23633 (2016)
17. Yang, H.X., Tian, L.: Enhancement of cooperation through conformity-driven reproductive ability. Chaos Solitons Fractals **103**, 159–162 (2017)
18. Javarone, M.A., Antonioni, A., Caravelli, F.: Conformity-driven agents support ordered phases in the spatial public goods game. EPL (Europhys. Lett.) **114**(3), 38001 (2016)

19. Niu, Z., Xu, J., Dai, D., et al.: Rational conformity behavior can promote cooperation in the prisoner's dilemma game. Chaos Solitons Fractals **112**, 92–96 (2018)
20. Kai, Y., Changwei, H., Qionglin, D., et al.: The effects of attribute persistence on cooperation in evolutionary games. Chaos Solitons Fractals **115**, 23–28 (2018)
21. Szabó, G., Tőke, C.: Evolutionary prisoner's dilemma game on a square lattice. Phys. Rev. E **58**(1), 69–73 (1998)

Effect of Pollution on Transboundary River Water Trade

Anand Abraham[✉] and Parthasarathy Ramachandran

Department of Management Studies, Indian Institute of Science,
Bangalore 560012, India
anandjacobabraham@gmail.com, parthar@iisc.ac.in

Abstract. Transboundary rivers are often polluted by multiple agents located along the river. In this work, we attempt to investigate the impact of pollution on bilateral river water trade. Specifically, we look at how trade quantity and price are affected by pollution damages. The contributions of the work are threefold. First, we propose a stylised two agent model of a polluted which describes allocation decisions. Next, we characterise the non-cooperative behaviour of this system of two agents. As the final result of this article we derive the participation condition for bilateral trade and an expression for the extent of trade which can thus happen.

Keywords: Conflict management · River sharing problem · River pollution · Bilateral trade

1 Introduction

Scarcity of freshwater has become a major cause of concern in past few decades in many parts of the world. In addition to scarcity, the problem of ill defined ownership rights threatens the water security in many of the shared river basins across the world. Increasing demand and overpopulation have triggered and accelerated conflicts among the users. In light of these competing uses, the scope of international watercourse agreements which was traditionally limited to navigation was expanded to include allocation in the nineteenth century [15]. The Helsinki rules of 1966 further expanded this scope to include water quality in order address concerns of pollution due to increased economic activity. However, traditional river sharing literature does not consider water quality when prescribing stable allocations [4,19,21]. The objective of this article is to frame a mathematical model for transboundary river water allocation which accommodates pollution and examine the conditions for cooperation.

Riparian states which are located on a transboundary river rely on international watercourse principles in order to assign property rights. The most commonly quoted principles are the Absolute Territorial Sovereignty (ATS) doctrine and the Unlimited Territorial Integrity (UTI) doctrine. The ATS doctrine results in a non-cooperative allocation of the resource which is both unfair [19]

© Springer Nature Switzerland AG 2019
D. C. Morais et al. (Eds.): GDN 2019, LNBIP 351, pp. 146–159, 2019.
https://doi.org/10.1007/978-3-030-21711-2_12

and inefficient [2,14]. The UTI doctrine which allocates all of the upstream water to a downstream country is not enforceable [6] and is infeasible if all the participants subscribe to this sharing rule [4]. Over the years, several alternative doctrines have been proposed as a middle ground between these two opposing doctrines. The most popular doctrine of this kind is the principle of reasonable and equitable utilisation which considers economical and hydrological factors to strike a balance between ATS and UTI doctrines. The theory of reasonable and equitable utilisation is based on the principle of limited territorial sovereignty. The introduction of such limits ensure long term beneficial utilisation of the river water without imposing significant harm to the other riparians. Unlike the ATS based allocation which can be self enforced under non-cooperation, an allocation in which water is transferred downstream requires non-water transfers (either monetary transfers or linked issue transfers) in order to bring about stability [6]. In literature, both cooperative game theoretic methods and market based methods are popular solution procedures used in the study of stable cooperative allocations in a transboundary river basin [11]. The cooperative game theoretic models use international watercourse principles to propose methods of sharing the benefits of cooperation. A popular sharing rule of this kind is the downstream incremental solution which is the unique distribution that satisfies the core lower bounds and aspirational upper bounds in a multi agent setting with quasi linear utilities [4]. The amenability of this sharing rule for single peaked utility functions were carried out by Ambec and Ehlers [3] who concluded that the non-cooperative core is non-empty and consist of the downstream incremental allocation where as the cooperative core could be empty for a river with three or more riparian states. Another sharing rule which is found to be appealing is the upstream incremental distribution which remains robust in the presence of reduced flows [2]. All these cooperative game theoretic models described ensure stability by prescribing an allocation of welfare which is in the core of the transferable utility game. The market based methods approach water as an economic commodity which can be transferred voluntarily between riparian states by means of bilateral trade. Investigating the conditions for such a bilateral trade in turn describes the conditions for a non-empty core. The price at which the trade takes place dictates the welfare allocation vector. Some of the early river sharing literature which used market based methods were restricted to two agents [11,17]. Wang extended the bilateral trade mechanism to propose a downstream bilateral distribution for a multi agent problem [21]. In addition to the cooperative game theoretic solutions and market based solutions, there are also solutions that does not rely on transferable utilities [7,19]. These methods propose a "fair" allocation of water based on bankruptcy methods [7,20] or by means of solution concepts like Shapley value [19]. In this work, we shall use a market based method to allocate water in a polluted river.

Due to ill defined ownership, responsibilities over externalities are also a cause of conflict in a transboundary river basin. The major externality that is studied in literature is pollution. Though pollution have not be modelled together with allocation, abatement of pollution and division of cleaning cost have been well discussed in river sharing literature. To divide clean up costs and decide abatement, both cooperative game theoretic solutions derived from international watercourse principles [9,10,16,18] and taxation based solutions for cooperation [1,13] have been studied in river sharing literature. These studies assume that allocation is not a decision which is affected by pollution. However, a negative externality like pollution also influences allocation decisions [12]. We notice that there exists a gap in the literature in terms of theoretical models that describe the economic behaviour of agents who share a polluted river. In this work, we consider the river water allocation decisions in the presence of pollutants and thereby address this gap in literature.

The rest of this paper is organised as follows. In Sect. 2, we introduce the two agent river sharing problem with problem under pollution externalities. Following that, in Sect. 3, we study the non-cooperative interaction. In Sect. 4, we use a market based method to induce cooperation between two agents along the river.

2   The Two Agent Model of a Polluted River

Consider a river system of two agents represented by the set $N = \{i, k\}$ where agent i is located upstream of agent k on a successive transboundary river[1]. Both these agents i and k receive annual endowments amounting to s_i and s_k respectively. The amount of water consumed by these agents are given by x_i and x_k respectively[2]. Upon consumption, the agents i and k generate benefits which are represented by the benefit functions $b_i(x_i)$ and $b_k(x_k)$ respectively. The following assumption characterises the benefit function.

Assumption 1 (Benefit function). *The benefit function $b_j(x_j) : \mathbb{R}_+ \to \mathbb{R}$ for $j \in N$ is non-negative function with a finite satiation point denoted as x'_j. Also, $b_j(0) \overset{\Delta}{=} 0$ and $b'_j(x_j) \to \infty$ as $x_j \to 0$. The benefit function is increasing for all $x_j < x'_j$ (i.e., $b'_j(x_j) > 0$) and it is decreasing for all $x_j > x'_j$ (i. e., $b'_j(x_j) < 0$). The benefit function is strictly concave with $b''_j(x_j) < 0\ \forall x_j \in \mathbb{R}_+$.*

The above mathematical assumption regarding the benefit function points to single peaked preferences for the agents. Models that consider quasi linear preferences [4,5,21] are typically more common and easy to treat compared to single peaked preferences [3]. Single peaked preferences are more realistic because it imposes an upper bound on benefit and represent satiable agents.

In this model, we make the following assumption regarding endowments.

---

[1] In order to focus our attention completely on pollution externality and its consequences on allocation, we restrict our study to two agents and do not consider complex river geographies. This approach simplifies our analysis and allows us to avoid excessive notation.

[2] The allocations x_i and x_k may be thought of as the net consumptions obtained after accounting for the return flows.

Assumption 2 (Upstream endowment). *The upstream endowment is greater than the agents satiation, i. e., $s_i > x_i'$.*

The above assumption makes sure that there is a flow of $s_i - x_i$ amount of water from agent i towards agent k. Such a non-zero flow is required for pollutant transfer along the river, the analysis of which is the substance of this study. The problem becomes uninteresting if the downstream agent k has enough availability of water. Therefore, we assume scarcity at the downstream agent's territory.

Assumption 3 (Downstream scarcity). *The downstream agent faces scarcity. The total available water for the downstream agent is lesser than its satiation level.*

The downstream scarcity assumption makes the problem into a river water allocation problem. This assumption ignores a scenario where downstream agent is satisfied with the quantity of water it gets but it still suffers from pollution damage. Such a scenario reduces to a cost sharing problem. The river pollution models [10,16] and pollution cost sharing models [9,18] are applicable under such a problem context. In the current analysis, we focus on how to allocate water under situations of conflict and cooperation among two riparian states in presence of pollution damages and hence choose to ignore such a scenario. The relevance of these above assumptions are further discussed in the next section where the equilibrium analysis of the non-cooperative behaviour is carried out.

The return flows as well as other effluent discharges could contain pollutants that cause welfare loss for the downstream agent. These pollutant effluents are discharged into the river in agent i's territory. The effect of pollution from agent i is manifested in the form of certain costs borne by the downstream agent k. If $q(x_i)$ represent the pollutant concentration expressed as a function of the upstream agent's consumption, the cost borne by agent k because of the pollution is captured by means of a damage function D_k. If the concentration of pollutant is $q(x_i)$ and d_p is cost per unit of pollutant consumed, then the amount of pollutant consumed upon extraction of x_k water by the upstream is $q(x_i) \times x_k$. In this work, we assume that both flow and pollutant concentration are accurately measurable and such a measurement is taken by agent k so as to facilitate its decision making process.

Assumption 4 (Damage function). *The damage cost is linearly proportional to the amount of pollutant consumed. i. e., $D_k = d_p \times q(x_i) \times x_k$.*

The damage function can represent a loss of welfare for agent k because of presence of pollutants. Alternatively, it may also be interpreted as the cost of cleaning purifying the water to appropriate quality standards[3].

Assumption 5 (Pollution concentration). *The pollution concentration is a positive real valued function of the flow $s_i - x_i$ and consequently the upstream*

---

[3] The assumption of linearity is taken for mathematical convenience as it allows for an easier understanding of the dynamics of the strategic situation.

consumption x_i. For an increase in the value of flow (decrease in the value of upstream consumption) the pollutant concentration goes down. Also, $q(x_i)$ at $x_i = s_i$ is defined as zero. i. e., $q(s_i) = 0$. Also, we make an assumption that the rate of change of concentration with change in upstream consumption is a constant positive number($\kappa > 0$). i .e.,

$$\frac{dq}{dx_i} = \kappa$$

Because of the assumption regarding pollution concentration, the pollutant concentration function can be thought of having the following format.

$$q(x_i) = \kappa x_i + q_0$$

where $q_0 > 0$ denotes the pollutant concentration at agent k's territory when agent i's consumption is zero.

Assumption 6 (Strategic nature). *Both agents are assumed to be rational and intelligent. The endowment values are assumed to be common knowledge. Moreover, the downstream agent has complete and perfect information regarding pollutant concentration $q(x_i)$ and flow $s_i - x_i'$.*

2.1   The Utility Function

The utility function for the upstream agent is constituted only by its benefit function $b_i(x_i)$.

$$u_i(x_i) = b_i(x_i) \tag{1}$$

In contrast, the utility function of the downstream agent k is affected by pollution damages. The utility function of the downstream agent is given by Eq. 2.

$$u_k(x_k) = b_k(x_k) - d_p q(x_i) \times x_k \tag{2}$$

2.2   Availability Constraint

The consumption of both players are constrained by their respective water availabilities. For agent i, the water availability is same as its endowment. This is represented below in constraint 3.

$$x_i \leq s_i \tag{3}$$

For agent k, the available water is the sum total of the agent's own endowment and flow from the upstream agent i. This translates to the mathematical expression 4.

$$x_k \leq s_k + s_i - x_i \tag{4}$$

The benefits, endowments, cost, and concentration components put together define a river sharing problem with pollution denoted by the 5-tuple $\langle N, s, b, d_p, q \rangle$. Where, N is the set of agents, s the set of endowments, b the set of benefits and q the pollution concentration function and d_p the damage cost per unit of pollutant consumed. A graphical representation of this stylized model is given in Fig. 1.

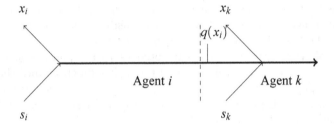

Fig. 1. Stylized two agent model for sharing a polluted river

3   Non-cooperative Equilibrium Analysis

The non-cooperative allocation decisions in the transboundary river basin is best representable using a sequential strategic interaction where agent i and k moves (decides on consumption) according to their geographical precedence (i.e., Agent i moves first and k follows). Such modelling of the strategic behaviour allows us to reason why the upstream agent always has at least a weak advantage in a transboundary river basin [19]. In addition, since agent k makes a fully informed choice regarding allocation it has to observe the consumption choice made by agent i. Since, the strategic interaction is sequential, we use the solution concept of Subgame Perfect Equilibrium (SGPE) to describe the non-cooperative behaviour of the two agent system.

Lemma 1. *For the river sharing problem with pollution denoted by $\langle N, s, b, d_p, q \rangle$, the consumption profile $(\hat{x}_i, \hat{x}_k)$ where*

$$\hat{x}_i = min(x'_i, s_i)$$

$$\hat{x}_k = min\left(\frac{db_k^{-1}}{dx_k}(d_p\kappa\hat{x}_i + d_pq_0), s_k + s_i - \hat{x}_i\right)$$

is a SGPE for the river sharing problem

Proof. We solve this game using backward induction. Let the upstream agent's consumption be some x_i. At this consumption, the pollutant concentration observed at the downstream agent is $q(x_i)$. A flow of water $s_i - x_i$ flows along the river towards agent i.

$$\text{Maximize } u_k(x_k) = b_k(x_k) - d_pq(x_i)x_k \tag{5}$$

$$\text{Subject to,} \quad x_k \le s_k + s_i - x_i \tag{6}$$

Note that this utility function is single peaked with a maximizer $\bar{x}_k$ which is characterised below using the first order condition for optimality.

$$b'_k(\bar{x}_k) = d_pq(x_i)$$

This means that, of all non-negative values of x_k, $\bar{x}_k$ would yield the highest utility to the downstream agent k if it were a feasible value. The feasibility of $\bar{x}_k$ depends on the availability of water (Eq. 6). i. e., if $\bar{x}_k \leq s_k + s_i - x_i$, then the optimal decision of agent k which is denoted as $\hat{x}_k$ is equal to $\bar{x}_k$. If however, $\bar{x}_k > s_k + s_i - x_i$, the increasing nature of the utility function until the point $\bar{x}_k$ ensures that $u_k(s_k + s_i - x_i) > u_k(x_k)$ $\forall x_k < s_k + s_i - x_i$. These two arguments are put together in Eq. 7 to describe the optimal decision of agent k given a consumption choice of x_i by agent i.

$$\hat{x}_k = min(\bar{x}_k, s_k + s_i - x_i) \tag{7}$$

The decision of agent k is conditional on the decision taken by agent i. The decision of agent i is not conditional on that taken by agent k. The decision problem of agent i is given below.

$$\text{Maximize } u_i(x_i) = b_i(x_i) \tag{8}$$
$$\text{Subject to,} \quad x_i \leq s_i \tag{9}$$

A similar set of arguments as the ones made for agent k can be made for agent i as well and the optimal consumption can be described by Eq. 10

$$\hat{x}_i = min(x_i', s_i) \tag{10}$$

If $q(\hat{x}_i)$ may be denoted as $\hat{q} = \kappa \hat{x}_i + q_0$ then a substitution can be made in Eq. 7 to get the equilibrium consumption level of agent k. Note that, pollution $q(x_i) = 0$ if $\hat{x}_i = x_i'$.

$$\hat{x}_k = min\left(\frac{db_k^{-1}}{dx_k}(d_p\kappa\hat{x}_i + d_p q_0), s_k + s_i - \hat{x}_i\right) \tag{11}$$

These consumption values in Eqs. 10 and 11 are obtained through backward induction and they induce a Nash equilibrium in each subgame of the sequential strategic interaction. This consumption profile is a Nash equilibrium which is thus sequentially rational and by definition is a Subgame Perfect Equilibrium (SGPE). □

Our assumptions regarding upstream endowment and downstream scarcity facilitates us to focus on only those cases where the upstream is satiated with $\hat{x}_i = x_i'$ and where a non-zero stock of pollutants flows downstream $q_0 > 0$. The downstream scarcity assumes that the endowments are such that there is some non-zero demand for water at agent k. This means the SGPE consumption $(\hat{x}_i, \hat{x}_k) = (x_i', s_k + s_i - x_i')$. In the later parts of this analysis, we assume this to be the non-cooperative consumption profile.

4   Cooperation in the Transboundary Basin

The non-cooperative consumption profile need not be efficient. Sharing of water is hence required to achieve efficiency in the basin. The inside options in a river sharing problem that can guarantee efficient solutions can be explored by trading water between agents along the river. Any such cooperative attempt can be broken down into a set of bilateral trades. Moreover, under a condition described in our problem setting where initial property rights are defined and transaction cost are negligible, then trade can bring about an efficient outcome [8]. In out problem setting, the initial property rights are defined asper the ATS doctrine which is naturally enforced under non-cooperation in a river. This non-cooperative equilibrium is unfair [19] and inefficient but it allocates the pollution cost to one agent. Under this problem setting which models two agents, the transaction costs can be assumed as zero. Wang [21] analyzes the cooperative behaviour of agents as the cumulative effect of bilateral trades which occur between consecutive agents. Under the assumptions of Wang's [21] model, trade occurs when downstream marginal benefits exceed the upstream marginal benefit. The trade stops under the equilibrium condition when the marginal benefits of both agents are equal. Similar dynamics can be observed in other market based methods [11,17].

A limitation of these studies is that the scope of cooperation is restricted to allocation decisions alone. Cooperation in a setting such as the one described in our study should also discuss how responsibilities over externalities in a river basin can be allocated. Like certain water claims, transboundary responsibilities have enforcement limitations. In the discussion on cooperation that follows, we look at a market based solution for water allocation and study how behaviour in the market is affected by the pollution variable and also look at how the proposed market based solution distributes welfare.

4.1 Dilution

During cooperation in a setting as described in this problem, the water quality at agent k which captured through the pollutant concentration, also changes as a function of allocations. Before formally characterizing the market based interaction, we revisit the welfare maximization problem of agent k and inspect how a change in allocation affects agent k's welfare. It was assumed earlier that the pollutant concentration observed in agent k's territory was an increasing function of agent i's consumption. So naturally a transfer of water in the downstream direction is expected to bring down the pollution concentration. This change in concentration of pollutants has its welfare implications also. These effects are examined in later after proposition 1 is introduced.

Let's say that after a transfer of x_{ik} amount of water in the downstream direction, the non-cooperative equilibrium consumption profile changes from $(\hat{x}_i, \hat{x}_k)$ to some $(x_i, x_k) = (\hat{x}_i - x_{ik}, \hat{x}_k + x_{ik})$. At this consumption profile, if a small positive quantity of water $\Delta x > 0$ is transferred downstream, in return for which a monetary transfer is made by agent k to agent i, the welfare of both the agents change. The welfare of both agents during the market interaction is the sum of welfare from consumption and transferable utility. If p is the price at which the water Δx is traded, then the expressions for the welfare of the agents i and k are given below.

$$z_i(x_i - \Delta x) = b_i(x_i - \Delta x) + \int_0^{x_{ik}} p \cdot dx + p\Delta x$$

$$z_k(x_k + \Delta x) = b_k(x_k + \Delta x) - d_p q(x_i - \Delta x)(x_k + \Delta x) - \int_0^{x_{ik}} p \cdot dx - p\Delta x$$

The condition under which both agents voluntarily agree to participate in such a trade is characterised by proposition 1.

Proposition 1. *At a consumption profile, (x_i, x_k), if both agents engage in trade over a very small positive quantity of water $\Delta x > 0$, then the following should hold true for a price value p*

$$b_i'(x_i) \le p \le b_k'(x_k) - d_p[\kappa(x_i - x_k) - q_0]$$

Proof. If agent k participates in trade, then its payoff from trade should exceed that from a situation without trade.

$$b_k(x_k) - d_p q(x_i) \le b_k(x_k + \Delta x) - d_p q(x_i - \Delta x)(x_k + \Delta x) - p\Delta x$$

On rearranging and simplifying

$$p \le \frac{b_k(x_k + \Delta x) - b_k(x_k)}{\Delta x} - d_p q(x_i - \Delta x) + d_p \frac{(q(x_i - \Delta x) - q(x_i))x_k}{\Delta x}$$

Applying the limit $\Delta x \to 0$ we get

$$p \le b_k'(x_k) - d_p q(x_i) + d_p \frac{dq}{dx_i} x_k$$

$$p \le b_k'(x_k) - d_p q(x_i) + d_p \kappa x_k \tag{12}$$

A similar argument can be made about agent i,

$$b_i(x_i) \le b_i(x_i - \Delta x) + p\Delta x$$

Rearranging and applying the limit, we get

$$p \ge b_i'(x_i) \tag{13}$$

From 12 and 13 we get,

$$b_i'(x_i) \le p \le b_k'(x_k) - d_p q(x_i) + d\kappa x_k$$

$$b_i'(x_i) \le p \le b_k'(x_k) - d_p[\kappa(x_i - x_k) + q_0]$$

At a consumption profile (x_i, x_k) trade occurs only if, there exist a non-empty set of price values which satisfies the above inequality. i. e., for trade

$$\{p \ni b_i'(x_i) \le p \le b_k'(x_k) - d_p[\kappa(x_i - x_k) + q_0]\} \ne \phi$$

$\square$

The above proposition also identifies the term $b'_k(x_k) - d_p q(x_i) + d\kappa x_k$ as the downstream agent's marginal willingness to pay. It can be seen that there occurs a cost reduction because of pollutant concentration change at the downstream state. If there were no such change in concentration, the damage would have been $d_p q(x_i)$. Due to the change in concentration, a reduction in costs (or savings) occurs. This is represented by $d\kappa x_k$ and in this work, this quantity is termed as *savings by dilution*. As one can observe, this savings depends on sensitivity of pollutant concentration to flow (which is represented through κ) and on the current allocation of agent k. At a higher allocation level at the downstream agent k, the agent k has a higher savings by dilution. The above proposition shows that in addition to change in welfare because of allocation, the downstream agent's welfare changes through a change in pollutant concentration also. This welfare change which depends on concentration change, rises with the amount of water allocated with the downstream agent.

In the light of proposition 1, we propose the following corollary and hence define an upper bound on the downstream's sensitivity to pollution (which is captured by d_p).

Corollary 1. *If the agents i and k were to initiate a trade process, the downstream agent's per unit damage has to satisfy the following inequality*

$$d_p \leq \frac{b'_k(s_k + s_i - x'_i)}{q_0 - \kappa(s_i + s_k - 2x'_i)}$$

if $q_0 - \kappa(s_i + s_k - 2x'_i) > 0$[4]

Proof. At a consumption profile $(\hat{x}_i, \hat{x}_k) = (x'_i, s_k + s_i - x'_i)$. For this consumption profile, the proposition 1 translates to the following inequality.

$$0 \leq b'_k(s_k + s_i - x'_i) - d_p \hat{q} + d_p \kappa(s_k + s_i - x'_i)$$

Upon rearranging these terms, we have

$$d_p \leq \frac{b'_k(s_k + s_i - x'_i)}{\hat{q} - \kappa(s_k + s_i - x'_i)}$$

Also, we know $\hat{q} = \kappa x'_i + q_0$. Substituting in the above equation, we get,

$$d_p \leq \frac{b'_k(s_k + s_i - x'_i)}{q_0 - \kappa(s_i + s_k - 2x'_i)}$$

$\square$

In the market based interaction, the trade will initiate only if the corollary 1 holds true. Trade will continue so far as proposition 1 holds true and trade ends at an equilibrium point and x^* amount of water is traded in the process. This equilibrium point is characterised in Eq. 14. x^* can be obtained by solving the Eq. 14.

$$b'_i(\hat{x}_i - x^*) = b'_k(\hat{x}_k + x^*) - d_p[\kappa(\hat{x}_i - \hat{x}_k - 2x^*) + q_0] \tag{14}$$

---

[4] Had $q_0 - \kappa(s_i + s_k - 2x'_i) < 0$, then the inequality would be $d_p \geq \frac{b'_k(s_k + s_i - x'_i)}{q_0 - \kappa(s_i + s_k - 2x'_i)}$.

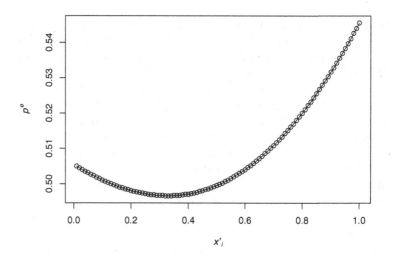

Fig. 2. Price (p^e) vs upstream consumption (x'_i)

4.2   Price of Water

The occurrence of trade a consumption profile (x_i, x_k) is characterised through proposition 1. The validity of the proposition depends on the existence of a price value $p > 0$ which satisfies

$$b'_i(x_i) \leq p \leq b'_k(x_k) - d_p q(x_i) + d\kappa x_k \tag{15}$$

If a free market behaviour was in place, the most natural price value to pick would be the equilibrium price which corresponds to the marginal welfare value at the market equilibrium. It is given by p^e,

$$p^e = b'_i(\hat{x}_i - x^*) = b'_k(\hat{x}_k + x^*) - d_p \left[\kappa(\hat{x}_i - \hat{x}_k - 2x^*) + q_0\right]$$

5   Numerical Example

Consider a benefit functions of the type $b_j(x_j) = B_j x_j \left(1 + log\left(\frac{A_j}{x_j}\right)\right)$ where $j \in \{i, k\}$. Consider the following values for the problem parameters

$$A_i = 1, A_k = 2, B_i = 1, B_k = 2$$

$$s_i = 1.2, s_k = 0$$

$$d_p = 0.5, \kappa = 0.5, q_0 = 0.1$$

The non-cooperative consumption profile for $(\hat{x}_i, \hat{x}_k) = (1, 0.2)$. At this consumption level, $\frac{b'_k(s_k + s_i - x'_i)}{q_0 - \kappa(s_i + s_k - 2x'_i)} = 2.3025 \geq d_p$ and therefore satisfies the corollary 1. This means that both agents posses an incentive to participate in the market mechanism. The traded quantity can be found by solving Eq. 14 and is found to be $x^* = 0.4205$. The equilibrium price in this context is found by using Eq. 15 and is found to be $p^e = 0.5455$.

5.1   Computational Study: Upstream Consumption and Price

In this computational study, we fix all parameters except A_i which is varied from 0.01 to 1 in steps of 0.01. At each of these values of A_i (which can also be identified as the satiation point for agent i), the equilibrium price is computed. The variation of equilibrium price with change in upstream consumption $\hat{x}_i = x_i'$ is shown in Fig. 2. It can be observed that the equilibrium price initially decreases when consumption increases. This trend continues until the price reaches some minimum level after which the price starts to rise.

This trend can be explained from the expression for the equilibrium price. The equilibrium price $p^e = b_k'(\hat{x}_k + x^*) - d_p[\kappa(\hat{x}_i - \hat{x}_k - 2x^*) + q_0]$. This expression may be rearranged to get the following form

$$p^e = b_k'(\hat{x}_k + x^*) - d_p[\kappa(x_i' - \hat{x}_k) + q_0] + 2d_p\kappa x^*$$

It may be seen that there are three terms in this expression. A marginal benefit term $b_k'(\hat{x}_k + x^*)$, a damage term $d_p[\kappa(x_i' - \hat{x}_k) + q_0]$ and a dilution term $2d_p\kappa x^*$. All three of these terms are influenced by a change in x_i'. The marginal benefit term increases with increase in x_i'. This variation may be seen in Fig. 3. The damage term $d_p[\kappa(x_i' - \hat{x}_k) + q_0]$ is negative for small values of x_i' as $(x_i' - \hat{x}_k)$ is negative. With an increase in x_i' the damage term also increases. Owing to this, the net marginal benefit $b_k'(\hat{x}_k + x^*) - d_p[\kappa(x_i' - \hat{x}_k) + q_0]$ decreases with increase in x_i'. The change of net marginal benefit with x_i' is depicted in Fig. 4. The dilution term $2d_p\kappa x^*$ increases with x_i'. This increasing trend is due to the fact that more room for trading is available as x_i' becomes larger. The cumulative effect of the net marginal benefit and the dilution term results in the cup shaped price curve. From this numerical example, it may be seen that the price is high for high values of upstream satiation. This behaviour is due to dilution. This leads one to believe that when the upstream agent is more in need of water, the price from trade is expected to go up and the upstream agent would consider trade a promising endeavour.

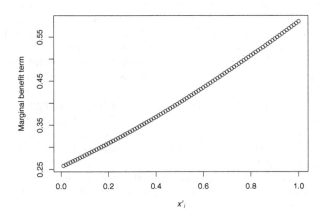

Fig. 3. Variation of the marginal benefit term $b_k'(\hat{x}_k + x^*)$ with x_i'

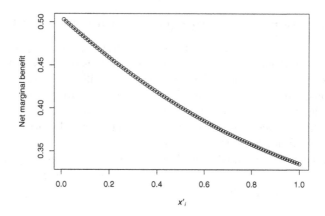

Fig. 4. Variation of the net marginal benefit $b'_k(\hat{x}_k + x^*) - d_p[\kappa(x'_i - \hat{x}_k) + q_0]$ with x'_i

6 Conclusion

In this work, we attempted to theoretically study the impact of pollution on river water allocation between riparian states during conflict and cooperation. The cooperation is induced by means of a market mechanism. It was observed that the amount of trade possible is limited by the pollution concentration in the water. Since, a greater level of trade quantity is profitable for both the agents, it is in their best interest for the agents to engage in abatement efforts to reduce pollution damages.

References

1. Alcalde-Unzu, J., Gómez-Rúa, M., Molis, E.: Sharing the costs of cleaning a river: the upstream responsibility rule. Games Econ. Behav. **90**, 134–150 (2015). http://www.sciencedirect.com/science/article/pii/S0899825615000305
2. Ambec, S., Dinar, A., McKinney, D.: Water sharing agreements sustainable to reduced flows. J. Environ. Econ. Manag. **66**(3), 639–655 (2013). http://www.sciencedirect.com/science/article/pii/S0095069613000429
3. Ambec, S., Ehlers, L.: Sharing a river among satiable agents. Games Econ. Behav. **64**(1), 35–50 (2008). http://www.sciencedirect.com/science/article/pii/S0899825607001674
4. Ambec, S., Sprumont, Y.: Sharing a river. J. Econ. Theory **107**(2), 453–462 (2002). http://www.sciencedirect.com/science/article/pii/S0022053101929497
5. Ansink, E., Houba, H.: Sustainable agreements on stochastic river flow. Resour. Energy Econ. **44**, 92–117 (2016). http://www.sciencedirect.com/science/article/pii/S0928765516000221
6. Ansink, E., Ruijs, A.: Climate change and the stability of water allocation agreements. Environ. Resour. Econ. **41**(2), 249–266 (2008). https://doi.org/10.1007/s10640-008-9190-3
7. Ansink, E., Weikard, H.P.: Sequential sharing rules for river sharing problems. Soc. Choice Welf. **38**(2), 187–210 (2012). https://doi.org/10.1007/s00355-010-0525-y

8. Coase, R.H.: The problem of social cost. J. Law Econ. **3**, 1–44 (1960). https://doi.org/10.1086/466560
9. Dong, B., Ni, D., Wang, Y.: Sharing a polluted river network. Environ. Resour. Econ. **53**(3), 367–387 (2012). https://doi.org/10.1007/s10640-012-9566-2
10. Gengenbach, M.F., Weikard, H.P., Ansink, E.: Cleaning a river: an analysis of voluntary joint action. Nat. Resour. Model. **23**(4), 565–590 (2010). https://doi.org/10.1111/j.1939-7445.2010.00074.x
11. Giannias, D.A., Lekakis, J.N.: Fresh surface water resource allocation between Bulgaria and Greece. Environ. Resour. Econ. **8**(4), 473–483 (1996). https://doi.org/10.1007/BF00357415
12. Giordano, M.A.: Managing the quality of international rivers: global principles and basin practice. Nat. Resour. J. **43**(1), 111–136 (2003)
13. Gomez-Rua, M.: Sharing a polluted river through environmental taxes. SERIEs **4**(2), 137–153 (2013)
14. Hardin, G.: The tragedy of the commons. Science **162**(3859), 1243–1248 (1968). http://science.sciencemag.org/content/162/3859/1243
15. Just, R.E., Netanyahu, S. (eds.): Conflict and Cooperation on Trans-Boundary Water Resources Conflict and Cooperation on Trans-Boundary Water Resources. Natural Resource Management and Policy, vol. 11. Springer, New York (1998). https://doi.org/10.1007/978-1-4615-5649-7
16. van der Laan, G., Moes, N.: Collective decision making in an international river pollution model. Nat. Resour. Model. **29**, 374–399 (2016)
17. Lekakis, N.J.: Bilateral monopoly: a market for intercountry river water allocation. Environ. Manag. **22**(1), 1–8 (1998). https://doi.org/10.1007/s002679900079
18. Ni, D., Wang, Y.: Sharing a polluted river. Games Econ. Behav. **60**(1), 176–186 (2007). http://www.sciencedirect.com/science/article/pii/S0899825606001412
19. Osório, A.: A sequential allocation problem: the asymptotic distribution of resources. Group Decis. Negot. **26**(2), 357–377 (2017). https://doi.org/10.1007/s10726-016-9489-3
20. Sechi, G.M., Zucca, R.: Water resource allocation in critical scarcity conditions: a bankruptcy game approach. Water Resour. Manag. **29**(2), 541–555 (2015). https://doi.org/10.1007/s11269-014-0786-9
21. Wang, Y.: Trading water along a river. Math. Soc. Sci. **61**(2), 124–130 (2011). http://www.sciencedirect.com/science/article/pii/S0165489610000946

Behavioral OR

Cognitive Style and the Expectations Towards the Preference Representation in Decision Support Systems

Ewa Roszkowska[1](✉) ⓘ and Tomasz Wachowicz[2] ⓘ

[1] University of Bialystok, Warszawska 63, 15-062 Bialystok, Poland
e.roszkowska@uwb.edu.pl
[2] University of Economics in Katowice, ul. 1 Maja 50, 40-287 Katowice, Poland
tomasz.wachowicz@uekat.pl

Abstract. The goal of this paper is to investigate how the cognitive style is related to the expectations towards the support mechanisms offered in the decision support system while analyzing the multiple criteria decision making problem. We analyze the decision makers' expectations regarding the forms of representing the results by the system (e.g. rankings vs. ratings), as well as the different ways in which they could declare their preferences (e.g. using numbers, words or pictograms). The relationship between the cognitive style determined by the Rational-Experiential Inventory and the decision makers declarations are examined using the correspondence and cluster analysis and fraction tests. The results to some extent confirm the postulates of the behavioral theory of decision making that the rational decision makers prefer the preference mechanisms that are more based on the numerical categories, oppositely to the experiential ones. Unfortunately, there are no clear patterns of preferences for versatile or indifferent decision makers. These results, however, do not so evidently correlate with the final recommendations of decision aiding methods.

Keywords: Decision support system · Cognitive profile · Preference analysis · Multiple criteria decision aiding

1 Introduction

In vast majority of real-life situations, making decisions requires a thorough and time-consuming trade-off analysis of the performances of many alternatives with respect to many and usually conflicting criteria. Therefore, a variety of multiple criteria decision aiding (MCDA) methods are proposed to support decision maker (DM) in their decision analyses [11, 32]. Various MCDA methods differ in the philosophy of preference elicitation, the aggregation algorithms they use, and the information their require from DMs [30], which often makes it difficult to choose the one that will be suitable to support DM in a particular decision making problem. Therefore, some researchers even propose the guidelines for selecting an appropriate MCDA technique, depending on the type and context of the decision making problem or the requirements imposed on the process of the preference elicitation [13, 16, 27], to make sure that the preference analyses and the final results will be sound and reliable.

© Springer Nature Switzerland AG 2019
D. C. Morais et al. (Eds.): GDN 2019, LNBIP 351, pp. 163–177, 2019.
https://doi.org/10.1007/978-3-030-21711-2_13

There are, however, some other factors, such as the behavioral ones, that may affect the results the MCDA methods produce [21, 25]. Some of them are related to the decision making skills and cognitive abilities of DMs, i.e. to the thinking/cognitive style and the way of analyzing of the decision-making processes [8, 10]. DMs that think fast and do not analyze the facts thoroughly are more prone to make some information processing errors that result from using intuition and heuristics [12, 33]. Consequently, they may misuse the decision support tools and produce false conclusions based on the recommendations provided by the decision support systems. Some empirical results show the potential relationship between cognitive capabilities and the decision making process and its results [7, 22]. Therefore, it seems vital to analyze, what decision support mechanisms fit the thinking styles the best (i.e. do not require an effort higher than the cognitive capabilities of DMs), and hence offer the facilitation that increases the chances for more accurate preference elicitation and generate final recommendations that are reflecting the intrinsic preferences of DMs best.

However, understanding the relationships between information processing style and the preferable decision aiding tools may be biased by the perspective used in describing the nature of style constructs themselves [10, 28]. Depending on which approach is used, i.e. unitary (e.g. parallel-competitive) or dual (i.e. orthogonal, e.g. default-interventionist) one, the conclusions on how the style affects the decision making effects may be different. In the former approach, the conclusions may be formulated based on the single index describing the single bipolar rationality-intuition dimension, e.g. the highly rational DMs are perceived simultaneously as little intuitive. In the second approach, both constructs form separate dimensions and hence the style is perceived as a mix of various intensities of both constructs, e.g. DM can be considered as both highly rational and intuitive.

Taking the above issues into account, the goal of this paper is twofold. First, we aim at investigating how the cognitive style affects: (1) the DM's expectations towards the form of the support mechanisms offered in the decision support system while analyzing the multiple criteria decision making problem, and (2) their actual selection of the MCDA method as most useful in solving real-world decision making problems. In particular, we analyze the DMs opinion regarding the most preferable and efficient way of preference representation that can be offered in the support system in quantitative or qualitative way; e.g. by means of numbers, linguistic terms, verbal descriptions or pictograms, and confront these opinions with the final choice of the decision support mechanisms that they recommend. Second, we wish to verify if the dual/orthogonal approach for measuring the cognitive style allows for better (more detail) description of the opinions and choices made by the DMs when evaluating the MCDA techniques.

In our analyses we use the results of the experiment conducted in online survey system (OSS). In the experiment, the decision making problem was predefined and three selected MCDA methods were implemented to support DMs, namely: AHP, SMART and TOPSIS [26]. Having completed the decision making phase, the respondents evaluated the OSS itself as well as the methods and their interfaces and provided the opinion regarding the most preferable, adequate and informative design of decision support mechanism. To describe cognitive profile of the participants, the Rational-Experiential Inventory (REI) test was used [17], which allows to identify two dimensions of such a profile: rationality and experientiality [8]. To identify the relationship between the cognitive style and the expectations regarding the most preferable way of support the correspondence analysis was used, which maps the relationships from multi-dimensional matrix into a two-dimensional space (using the aggregation that keeps as much of original information as possible) and hence allows to analyze them based on the notion of distances.

The rest of the paper is organized as follows. In the Sect. 2 the behavioral aspects of decision support are discussed and research questions asked. In Sect. 3 we described the experiment we had conducted, while in Sect. 4 the results are presented. We summarize the results in Conclusions and answer the research questions.

2 Behavioral Factors and Decision Support

2.1 Investigating an Impact of Thinking Styles on Decision Process

There were early works of Simon, Kahneman and Tversky that paid attention to the behavioral issues of decision making processes and the cognitive limitation of DMs [29, 33]. The postulates of including some notions of behavioral analysis into the process of verifying the effects of decision support and the usefulness of decision aiding tools were and still are raised by the MCDM methodologists [21, 25, 34]. Some experimental works show that various behavioral elements may affect the decision making and decision support, among others the cognitive style of DMs, which is defined as the consistent individual differences in preferred ways of organizing and processing information and experience [24].

Green and Hughes [14] experimentally confirm the existence of interaction between the cognitive style and the type of training, which affected the decision maker initial use of a decision support system (DSS). The manager's cognitive style was measured by them by means of Myers-Briggs Type Indicator (MBTI) [3].

Engin and Vetschera [7] experimentally studied the relationship between the cognitive style and decision quality when using tabular or graphical representations of information. The error rate, which measured the discrepancy between a reference ranking of alternatives and the one built by the DM, seemed to decrease when the analytical orientation of DM's increased for tabular representation, while it increased for graphical one. Moreover, the effect of cognitive style was stronger for tables than for graphical representation. To described an individual's position they used competitive analytic-intuitive scale measure by Cognitive Style Index [2].

Lu et al. [22] analyzed the effects of cognitive style (measured by means of MBTI) and type of decision support model on the decision support acceptance. Three different support models were considered: the fuzzy weighted-sum model, Saaty's analytic hierarchy process, and the linear weighted-sum model. They inconclusively observed, that the cognitive style allowed to described the relationships among different acceptance measure only for one of these models (the fuzzy weighted-sum one), but not for others.

Chakraborty et al. [5], on the other hand, examined the DMs' acceptance of new technology using Technology Acceptance Model [6]. They showed that cognitive style has significant direct effects on perceived usefulness, perceived ease of use, and subjective norms, and both perceived usefulness and subjective norms affect actual technology usage. The individual's cognitive style was measured by means on Kirton Adaption–Innovation Inventory [4], which is an instrument for measuring the style on an adaptor–innovator continuum.

All these works show that cognitive style can affect the decision making process and results, however, the cognitive style itself can be differently perceived and interpreted. Thus, the question arises, what can be the simplest yet sufficiently informative way of measuring the cognitive style.

2.2   Rational/Analytical and Experiential/Intuitive Thinking Modes

There is an extensive discussion among the psychologists on how to define the cognitive style best. An approach based on the dual-processing distinction seems to be most grounded among the researchers, and it stems from the early works in a field of psychology and decision making [33, 36]. It makes a distinction between two thinking modes: rational/analytical and experiential/intuitive ones. For the terminology issues Stanovich simply calls them System 1 and System 2 [31], but we will use these terms interchangeably.

Within the cognitive psychology literature, it has been suggested that rationality and intuition are two coexisting information-processing systems, however there is lack of consensus about the theoretical relation between them [2, 9]. As was pointed out [35]: "models of individual differences in cognition differ as to whether intuition and analysis are viewed as bipolar opposites or as two independent unipolar dimensions. The distinction concerns whether one can be as follows: (i) either intuitive or analytical or (ii) both intuitive and analytical in orientation. The first implies a negative relation between the constructs, whereas the second implies no relation between intuition and analysis".

The cognitive style, that is an orthogonal mix of the rational and intuitive approaches, can be determined by the Rational-Experiential Inventory (REI) [8]. This is a psychometric test consisting of the series of questions regarding the way of thinking and reasoning that the responder usually implements in their everyday life. However, some experimental results indicate various problems with understanding a 40-item original inventory [20]. Hence, some modifications of original REI test are proposed, such as a shorten version of REI test that consists of 20 items, i.e. the REI-A test [23].

According to REI results and some aggregation principles, the DMs can be assigned into one of four classes that differ in the combination of scores for rational and experimental modes [1, 18]. Those highly rational and highly experiential (HRHE) are called *cognitively versatile*, and are considered to have the skills to consider the problems both in details and as a big picture, when required. DMs from class HRLE are *detail conscious* (or *rational*) and have tendency to approach problem step by step in systematic way, while in contrast those from the class LRHE are *big picture conscious* (or *experiential*) use mostly intuition and are able to detect emergent issue ahead. Finally, those from the LRLE class (*non-discerning* or *indifferent*) seems not to be willing to engage their own cognitive resources in information processing (neither analyze nor base on the intuition) but rather rely on the opinions of others.

2.3 Research Questions

Taken into account all the issues raised in the previous two subsections some research questions may be formulated. First, following the extensive discussion regarding the competing approaches to the definition of the cognitive style [10] we would like to answer the following question:

Q1: Does the orthogonal definition of the cognitive styles allow to define the cognitive style classes that differ significantly in DMs' most preferred ways of preference representation and declaration (preference representation schema)?

Taking into account earlier studies an experiments (see e.g. [19]), *the detail conscious* DMs may be perceived as rational in bilateral perspective, and we may expect that they prefer more quantitative ways of preference representations. Similarly, *the big-picture conscious* ones can be called experimental in bilateral perspective and hence they would probably prefer to operate with non-numerical representation of preferences. We will try to confirm these two theses, but we are more interested in finding:

Q2: What are the most preferred preference representation schemas by *versatile* or *indifferent* DMs?

To answer Q1 and Q2 we will use the dataset from the decision making experiment, in which the REI test was implemented. We will cluster the DMs into four groups as defined in Sect. 2.2, and analyze the answers they gave in the post-task questionnaires regarding: the way in which they wish to declare the preferences the most; and the most preferable representation of the results of the decision making process and the evaluation of alternatives. These are however, the self-reported general declarations and we would like to confront them with the DMs' final recommendations regarding the most preferable decision support method, i.e. one of three different ones that they used in the experiment. Therefore, we ask:

Q3: Do the cognitive style and the corresponding preference representation schema correlate with the choice of the best decision aiding method?

This would allow us to check whether DMs are truly interested in using the decision aiding tool that fits their cognitive capabilities the most, or they would rather opt, for instance, for a method that is quick, less time-consuming or has a nicer user interface.

3 Decision Making Experiment

To find the answers for the research questions posed in Sect. 2.2, we organized the decision making experiment in OSS. In the experiment the hypothetical problem of choosing a flat to rent was consisted for which five predefined alternatives were defined, each describing the resolution levels for five evaluation criteria. Since the participants were 413 students of four Polish universities, the problem was stylized to their decision making context. Table 1 presents the full decision matrix of the problem under consideration.

Table 1. Decision matrix in OSS experiment

Alternative	Rental costs	No. of rooms	Size	Equipment	Travel time to university
A	950 PLN	2 rooms (including 1 room with a kitchenette)	35 m^2	Fridge, washing machine, microwave	10–12 min
B	1200 PLN	3 rooms (including a living room with a kitchenette)	54 m^2	Fridge, washing machine, dishwasher, wireless internet	30–35 min
C	900 PLN	2 rooms + kitchen (separate)	35 m^2	Fridge, washing machine, cable internet	20–25 min
D	700 PLN	1 room + kitchen	25 m^2	Fridge, washing machine, TV, cable TV, cable internet	30–35 min
E	950 PLN	1 room + kitchen	54 m^2	Fridge, washing machine, cable internet	20–25 min

The experiment consisted of several steps that were related to the process of preference elicitation and decision support. At the beginning the respondents read the case and set an individual ranking of alternatives using the holistic approach. Using the boxes that visualized the alternatives and the drag-and-drop mechanism, they organized the boxes in an order that were supposed to reflect their individual subjective preferences (no instruction about the references were given to the participants).

In the next steps the decision analysis was conducted, which started from elicitation of criteria weights, where the participants used both AHP-based pair-wise mechanism and linguistic evaluation. Unsatisfied with the results produced by these two methods, they could also assign the weights directly themselves. Then the consequences of the alternatives were evaluated using three implemented MCDA methods: AHP, SMART and TOPSIS that differed in the preference elicitation schema. To each method the corresponding user interface was designed that was supposed to fit the cognitive requirements of the method itself. For AHP the sliders were used for each compared pair with accompanying verbal description of evaluation set by the slider. For SMART the tables were presented, which had to be filled by DMs directly with numbers representing their preferences. Since TOPSIS evaluates the quantitative criteria automatically using the notion of distances, there was a need for implementing a method for evaluation of two qualitative criteria in our problem, namely no. of rooms and equipment. Here, the DMs declared the preferences using pictograms, i.e. for each option seven empty stars were assigned and DMs colored in yellow as many of them as required to express the option performance. The screen-shots of the interface for each of MCDA method are presented in Fig. 1.

Finally, the rankings of alternatives obtained by means of these three MCDA methods were presented to the respondents, including the scores they obtained and their graphical representations (five stars rating for TOPSIS and SMART or colored circles for AHP), as shown in Fig. 2.

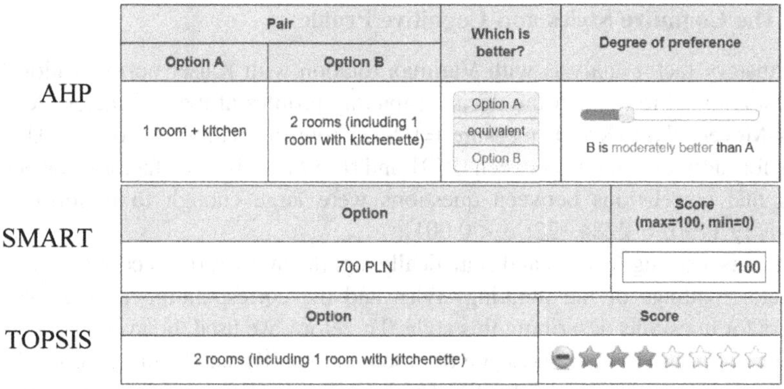

Fig. 1. User interface used for preference declarations in each MCDM method (Color figure online)

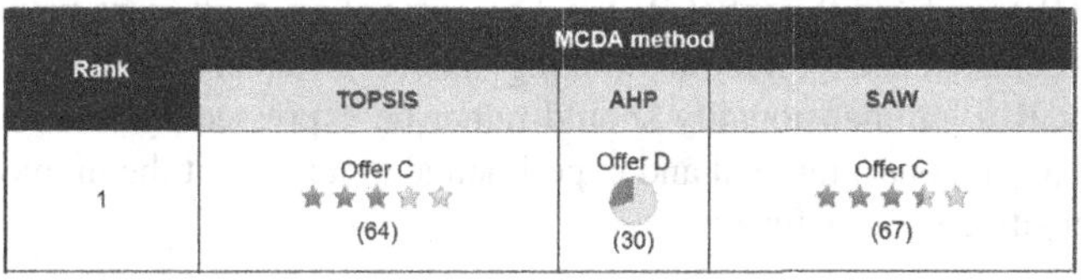

Fig. 2. Display of the results of MCDA process in OSS (Color figure online)

In the series of post-decision making questionnaires the respondents had evaluated the whole decision support process offered in OSS. They were asked to evaluate each support method, their ease of use, interface, reliability etc. We also asked about their opinion regarding some aspects related to the optimal design of the decision support mechanism and software support, such as the way of representing the results, describing the alternatives scores in final rankings, and best possible way of declaring the preferences in the preference elicitation process.

Finally, the respondents filled the REI test, which allowed us to determine their cognitive profiles and link them with their evaluation and expectations toward the decision support. As our respondents were non-native English speakers, we were afraid of misunderstanding problems related to some language nuances, hence in this study we used a shorten version of REI test that consists of 20 items, i.e. the REI-A test.

Correspondence analysis was used to describe the relationships among the factors under analysis [see e.g. 15].

4   Results

4.1   The Cognitive Styles and Cognitive Profiles

Confirmatory factor analysis with Varimax rotation with Kaiser normalization for the REI test allowed to defined the decision-making profiles at the satisfactory level. The Kaiser-Meyer-Olkin (KMO) measure (=0.852) confirms sample adequacy. The KMO values for individual items between 0.791 and 0.887 are also satisfactory. Bartlett's test shows that correlations between questions were large enough to perform a factor analysis $[\chi^2(190) = 2284.122; p < 0.001]$.

There is a strong (>0.99) and statistically significant ($p < 0.01$) correlation between the factor loadings of the thinking styles and the corresponding average values of answers for questions describing this style. Therefore, we used the average values from questions 1–10 as the scalar measure of rational mode (R), and from questions 11–20 to describe their experiential mode (E). These average values were used to classify the respondents within each mode into two classes: L (less or equal to average) and H (above the average). The combination of these two classes for two modes make four different categories of cognitive profiles: Versatile (HRHE), Rational (HRLE), Experiential (LRHE), and Indifferent (LRLE). The correlation coefficient between average values of rational and experiential thinking modes is equal −0.054. This is a first indicator that REI's dimensionality should rather be expressed by two interacting but independent (orthogonal) rational and experimental systems, not the unimodal one, and suggests a positive answer for Q1.

4.2   Expectations Towards the Declaration of Preferences

First, we analyzed the respondent's choices regarding the most preferred way of defining their preferences in the preference elicitation process. The numbers and fractions of choices are shown in Table 2 and visualized in Fig. 3a.

Taking into account the relatively low fractions in choosing the forms of preference declarations other than three most frequently chosen classes we removed the latter form the correspondence analysis to make the visualization more readable.

Table 2. Decision making profiles vs preferred forms of preference impartation (N = 413).

Declaration of preferences	Indifferent	Versatile	Rational	Experiential
Numerical	60 (56.1%)	54 (61.4%)	78 (69.6%)	62 (58.5%)
Pictorial	30 (28.0%)	29 (32.9%)	22 (19.6%)	34 (32.1%)
Verbal	17 (15.9%)	5 (5.7%)	8 (7.2%)	9 (8.5%)
In other way	0 (0%)	0 (0.0%)	4 (3.6%)	1 (0.9%)
Sum	107 (100%)	88 (100%)	112 (100%)	106 (100%)

It seems that the Versatile and Experiential DMs are relatively close to each other (fractions in Table 2 are quite similar), but far from Rational and Indifferent. Among all the classes most DMs choose numerical, next pictorial and finally verbal description of preferences. This is a dominant choice (69.6%) for DMs with Rational profile. However, Experiential and Versatile DMs are also quite frequently choosing pictorial way of preference declaration, while the Indifferent ones choose the verbal declarations most frequently than others.

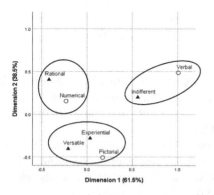 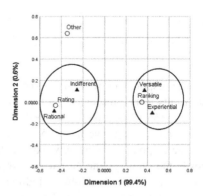

a) Cognitive profiles vs forms of preference declarations.

b) Cognitive profiles vs. forms of representing results by the system

Fig. 3. Correspondence analysis for cognitive profiles in OSS (1)

The differences in fractions for Rational profile and two other classes with low rationality index (Indifferent, Experiential) are significant ($p < 0.072$). The DMs with higher rationality index (Rational and Versatile) differ significantly in the choice of pictorial declarations ($p < 0.017$). Also, Indifferent and Experiential DMs differ significantly ($p < 0.05$) in choosing the verbal declarations among themselves. This is another premise to answer Q1 positively.

4.3 Expectation Towards the Representation of Final Results of Decision Analysis

Next, we have analyzed the DMs' expectations about the forms of the representation of the results obtained in the decision analysis phase. The fractions of user's expectations regarding the cardinal quantitative representation of the results (ratings) vs. ordinal representation by means of rankings only or other types of representations are presented in Table 3 and visualized by means of correspondence analysis in Fig. 3b.

Table 3. Decision profiles classes vs forms of representing results by the system (N = 413).

Forms of representing the results of decision analysis	Indifferent	Versatile	Rational	Experiential
The rating of each alternatives	49 (45.8%)	30 (34.1%)	56 (50.0%)	35 (33.0%)
Ordering (ranking) alternatives	55 (51.4%)	56 (63.6%)	53 (47.3%)	69 (65.1%)
I need other information	3 (2.8%)	2 (2.3%)	3 (2.7%)	2 (1.9%)
Sum	107 (100%)	88 (100%)	112 (100%)	106 (100%)

The respondents with high experiential index (HE: Versatile, Experiential) prefer the results to be presented in a form of ranking only. Their fractions of choosing the rankings (65.1% and 63.6% respectively) differ significantly from those determined for DMs with low intuitive index (51.4% and 47.3%, respectively).

The respondents with low experiential index (LE: Indifferent or Rational) prefer mostly the rating as the best way of representing the results of the decision analysis. There is as much as 45.8% of Indifferent DMs and 50.0% of Rational who choose rating, while only 34.1% of Versatile and 33.0% of Experiential do so.

The differences in choices between LE and HE are significant ($p < 0.048$), and this is in fact the only situation in our study, where unimodal approach to the definition of cognitive style could be used to describe the differences in sufficiently accurate way.

4.4 Expectations Towards the Representation of Offers' Evaluation

We have also analyzed the responses of the DMs for more detailed question regarding the most preferred representation of the alternatives evaluation in the final ranking. Various combinations of representations suggested by the respondents were clustered into three classes: pure numerical representation, non-numerical representation, or the mixed one that joins the advantages of all three forms. In Table 4 the numbers of respondents selecting each representation of alternatives' evaluations are provided and the results are visualized in Fig. 4a.

Table 4. Preferred forms of presentation of alternatives evaluation in the final ranking (N = 413).

Description of alternatives in final ranking	Indifferent	Versatile	Rational	Experiential
Only numerical	48 (44.8%)	37 (42.1%)	63 (56.2%)	48 (45.3%)
Mix with numerical	34 (31.8%)	34 (38.6%)	29 (25.9%)	34 (32.1%)
Non-numerical	25 (23.4%)	17 (19.3%)	20 (17.9%)	24 (22.6%)
Sum	107 (100%)	88 (100%)	112 (100%)	106 (100%)

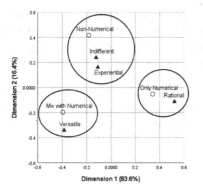 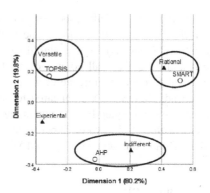

a) Cognitive profiles vs forms of prefer-   b) Cognitive profiles vs. recommended
ence declarations.   method

Fig. 4. Correspondence analysis for cognitive profiles in OSS (2)

Here we see that Indifferent and Experiential profiles are relatively close to each other (unlike to previous results), but far from Versatile and Rational. The DMs from all the classes prefer most the numerical evaluation, then mixed one and the least – the non-numerical representation of offers' evaluation. Rational DMs are the only ones to prefer numerical scores in majority (56.2%). The Versatile also like the numerical evaluations of offers but combined with some alternative representations (verbal and/or pictorial), while the DMs with low rationality index (Experiential, Indifferent) are more satisfied with the non-numerical representation of offers' values than DMs from other classes.

The differences in fraction of choices made by the Rational and Versatile DMs are significant ($p < 0.023$) for numerical and mixed representations. They similarly frequently choose non-numerical representation, but the fraction of such choices is relatively low (<19.3%). It seems, however, that Indifferent and Rational profiles differ significantly in choosing the numerical scores ($p < 0.05$).

4.5   Recommendation of MCDA Method

Finally, we have analyzed what are the recommendations towards using a specific MCDA method by the DMs from different classes. This way we wished to confront the DMs most preferred ways of handling the preferences with their real choices made as the DSS users, who were offered the software implementations of particular MCDA techniques. The results are presented in Table 5, and visualized in Fig. 4b.

Table 5. Recommended methods (N = 413).

Recommended method	Indifferent	Versatile	Rational	Experiential
AHP	35 (32.7%)	23 (26.2%)	30 (26.8%)	33 (31.1%)
SMART	31 (29.0%)	20 (22.7%)	38 (33.9%)	23 (21.7%)
TOPSIS	40 (37.4%)	42 (47.7%)	43 (38.4%)	49 (46.3%)
None of them	1 (0.9%)	3 (3.4%)	1 (0.9%)	1 (0.9%)
Sum	107 (100%)	88 (100%)	112 (100%)	106 (100%)

Surprisingly, TOPSIS appeared to be a choice of relative majority of DMs within each class of cognitive profiles. The highest percentage of TOPSIS choices was in Versatile (47.7%) and Experiential (46.3%) classes. This is an interesting finding, since the respondents from all cognitive profiles had previously preferred in majority the numerical forms of preference declarations (see Table 2).

Indeed, Rational DMs appeared to choose SMART more frequently than others, Versatile more frequently than others chose TOPSIS, and the Indifferent – AHP. The general profile of choices of Experimental DMs makes them similarly close to TOPSIS and AHP but definitely far form SMART.

5   Conclusions

In this study we tried to analyze what are the DMs' willingness and expectations for operating with various types of preference information during the process of decision analysis conducted in the decision support system. Some of our results confirm the general propositions formulated by the behavioral theory of decision making, that link the thinking styles with some schemas of DMs activities focused on solving the decision making problems. For instance, the Rational DMs prefer in majority the usage of ratings, numerical only description of alternatives and numerical declaration of preferences more frequently than respondents from other classes. They also prefer not to use the pictorial declaration preferences more than respondents from other classes of decision making profiles. Contrary, the Experiential ones do not process information extensively and hence need no precise data to be provided. This simply overlap with the general description of rational and experiential actors provided by Epstein [8].

However, trying to find the answers for the research questions we were able to shed a new light on some nuances of the potential impact of cognitive styles on the expectations and use of the decision support tools. It seems that the question Q1 can be answered positively, as the orthogonal definition of the cognitive styles was absolutely necessary to described the differences in DMs' most preferred ways of preference representation and declaration. When we look at the groups of similar profiles encircled at Figs. 3 and 4, the Rational and Experiential ones always constitute separate groups, while the Versatile and Indifferent classes, each independently, sometimes are more alike to Rational, and sometimes to the Experiential one. A good summary of our results, that allow to formulate the answer for question Q2 is presented in Table 6.

Table 6. Profiles and nearest choices according to correspondence analysis.

Choices regarding:	Indifferent	Versatile	Rational	Experiential
Preference declaration	Verbal	Pictorial	Numerical	Pictorial
Results representation	Rating	Ranking	Rating	Ranking
Offer evaluation	Non-numerical	Mixed with numerical	Only-numerical	Non-numerical
Recommended method	AHP	TOPSIS	SMART	TOPSIS/AHP

The Versatile on general seem to be more like Experimental in preference declaration and results representation (no quantitative representation required). However, they would like to obtain a little more specific information about differences in alternative performance that would be somehow linked to numbers. This matches with the support method they are closest to, which appears to be TOPSIS. The Indifferent, on the other hand, prefer to define their preferences verbally (like no other), and wish to have a cardinal information as the result (rating as Rational). Interestingly, they would like to receive such cardinal information about alternatives in non-numerical way (as Experiential). This explains their general choice of AHP.

The answer for question Q3 is not so evident. Looking at Table 6 one may say that some generalized outcomes provided by correspondence analysis allow to answer Q3 positively. Cognitive style and choices regarding preference representation match the methods the DMs recommend to use. Rational like numbers and hence choose SMART, while Experiential dislike them and they opt for either TOPSIS or AHP. However, if we look at the profiles of choices (Table 5), we will see that in each class the relative majority of DMs choose TOPSIS. If we confront it with another finding, that the absolute majority in each class prefers to declare preferences by means of numbers (Table 2), a complicated picture of some contradictory expectations and needs of DMs is given, that needs further and deeper investigation.

A need of more detailed research arises, that would identify the preferences towards using particular decision support algorithms accompanied by some additional visualization techniques, with a measure of the efficiency of their use. This could help to develop the support mechanisms that are fit to the cognitive capabilities of the decision makers, reduce the potential errors they could make using the heuristics typical to the profile of their thinking style, and provide them with a type of information that they are able to process and use efficiently to make the best decisions.

Acknowledgements. This research was supported with the grant from Polish National Science Centre (2016/21/B/HS4/01583).

References

1. Akinci, C., Sadler-Smith, E.: Assessing individual differences in experiential (intuitive) and rational (analytical) cognitive styles. Int. J. Sel. Assess. 21(2), 211–221 (2013)
2. Allinson, C.W., Hayes, J.: The cognitive style index: a measure of intuition-analysis for organizational research. J. Manag. Stud. 33(1), 119–135 (1996)
3. Briggs, K.C., Myers, I.B.: The Myers-Briggs Type Indicator: Form G. Consulting Psychologists Press, Palo Alto (1977)
4. Carne, G., Kirton, M.J.: Styles of creativity: test-score correlations between Kirton adaption-innovation inventory and Myers-Briggs type indicator. Psychol. Rep. 50(1), 31–36 (1982)
5. Chakraborty, I., Hu, P.J.-H., Cui, D.: Examining the effects of cognitive style in individuals' technology use decision making. Decis. Support Syst. 45(2), 228–241 (2008)
6. Davis, F.D.: Perceived usefulness, perceived ease of use, and user acceptance of information technology. MIS Q. 1, 319–340 (1989)
7. Engin, A., Vetschera, R.: Information representation in decision making: the impact of cognitive style and depletion effects. Decis. Support Syst. 103, 94–103 (2017)

8. Epstein, S., et al.: Individual differences in intuitive–experiential and analytical–rational thinking styles. J. Pers. Soc. Psychol. **71**(2), 390–405 (1996)
9. Evans, J.S.B.: In two minds: dual-process accounts of reasoning. Trends Cogn. Sci. **7**(10), 454–459 (2003)
10. Evans, J.S.B., Stanovich, K.E.: Dual-process theories of higher cognition: advancing the debate. Perspect. Psychol. Sci. **8**(3), 223–241 (2013)
11. Figuera, J., Greco, S., Ehrgott, M. (eds.): Multiple Criteria Decision Analysis: State of the Art Surveys. Springer, Boston (2005). https://doi.org/10.1007/b100605
12. Gilovich, T., Griffin, D., Kahneman, D.: Heuristics and Biases: The Psychology of Intuitive Judgment. Cambridge University Press, Cambridge (2002)
13. Górecka, D.: On the choice of method in multi-criteria decision aiding process concerning European projects. Mult. Criteria Decis. Making **6**, 81–103 (2011)
14. Green, G.I., Hughes, C.T.: Effects of decision support systems training and cognitive style on decision process attributes. J. Manag. Inf. Syst. **3**(2), 83–93 (1986)
15. Greenacre, M., Blasius, J.: Multiple Correspondence Analysis and Related Methods. Chapman and Hall/CRC, London (2006)
16. Guitouni, A., Martel, J.-M.: Tentative guidelines to help choosing an appropriate MCDA method. Eur. J. Oper. Res. **109**(2), 501–521 (1998)
17. Handley, S.J., Newstead, S.E., Wright, H.: Rational and experiential thinking: a study of the REI. Int. Perspect. Individ. Differ. **1**, 97–113 (2000)
18. Hodgkinson, G.P., Clarke, I.: Conceptual note: exploring the cognitive significance of organizational strategizing: a dual-process framework and research agenda. Hum. Relat. **60**(1), 243–255 (2007)
19. Kahneman, D., Tversky, A.: Choices, Values, and Frames. Cambridge University Press, Cambridge (2000)
20. Klaczynski, P.A., Fauth, J.M., Swanger, A.: Adolescent identity: rational vs. experiential processing, formal operations, and critical thinking beliefs. J. Youth Adolesc. **27**(2), 185–207 (1998)
21. Korhonen, P., Wallenius, J.: Behavioral issues in MCDM: neglected research questions. In: Clímaco, J. (ed.) Multicriteria Analysis, pp. 412–422. Springer, Heidelberg (1997). https://doi.org/10.1007/978-3-642-60667-0_39
22. Lu, H.-P., Yu, H.-J., Lu, S.S.: The effects of cognitive style and model type on DSS acceptance: an empirical study. Eur. J. Oper. Res. **131**(3), 649–663 (2001)
23. Marks, A.D., et al.: Assessing individual differences in adolescents' preference for rational and experiential cognition. Pers. Individ. Differ. **44**(1), 42–52 (2008)
24. Messick, S.: Individuality in Learning. Jossey-Bass, San Francisco (1976)
25. Morton, A., Fasolo, B.: Behavioural decision theory for multi-criteria decision analysis: a guided tour. J. Oper. Res. Soc. **60**(2), 268–275 (2009)
26. Roszkowska, E., Wachowicz, T.: Analyzing the applicability of selected MCDA methods for determining the reliable scoring systems. In: The 16th International Conference on Group Decision and Negotiation. Western Washington University, Bellingham (2016)
27. Saaty, T.L., Ergu, D.: When is a decision-making method trustworthy? Criteria for evaluating multi-criteria decision-making methods. Int. J. Inf. Technol. Decis. Making **14**(06), 1171–1187 (2015)
28. Sadler-Smith, E.: The intuitive style: relationships with local/global and verbal/visual styles, gender, and superstitious reasoning. Learn. Individ. Differ. **21**(3), 263–270 (2011)
29. Simon, H.A.: A behavioral model of rational choice. Q. J. Econ. **69**(1), 99–118 (1955)
30. Słowiński, R., Greco, S., Matarazzo, B.: Axiomatization of utility, outranking and decision rule preference models for multiple-criteria classification problems under partial inconsistency with the dominance principle. Control Cybern. **31**(4), 1005–1035 (2002)

31. Stanovich, K.E.: Who is Rational? Studies of Individual Differences in Reasoning. Psychology Press, Hove (1999)
32. Triantaphyllou, E.: Multi-criteria decision making methods. In: Triantaphyllou, E. (ed.) Multi-criteria Decision Making Methods: A Comparative Study, vol. 44, pp. 5–21. Springer, Boston (2000). https://doi.org/10.1007/978-1-4757-3157-6_2
33. Tversky, A., Kahneman, D.: Judgment under uncertainty: heuristics and biases. In: Wendt, D., Vlek, C. (eds.) Utility, Probability, And Human Decision Making, vol. 11, pp. 141–162. Springer, Dordrecht (1975). https://doi.org/10.1007/978-94-010-1834-0_8
34. Wallenius, J., et al.: Multiple criteria decision making, multiattribute utility theory: recent accomplishments and what lies ahead. Manag. Sci. **54**(7), 1336–1349 (2008)
35. Wang, Y., et al.: Meta-analytic investigations of the relation between intuition and analysis. J. Behav. Decis. Making **30**(1), 15–25 (2017)
36. Wason, P.C., Evans, J.S.B.: Dual processes in reasoning? Cognition **3**(2), 141–154 (1974)

Cue Usage Characteristics of Angry Negotiators in Distributive Electronic Negotiation

Sriram Venkiteswaran$^{(\boxtimes)}$ (ID) and R. P. Sundarraj (ID)

Indian Institute of Technology Madras, Chennai, India
sriramve2000@gmail.com, rpsundarraj@iitm.ac.in

Abstract. The role of anger in negotiation is explored in considerable depth in many papers in the literature. In the electronic negotiation situation, one way to express anger (in addition to plain textual messages) is through the use of emoticons and para-linguistic cues. Cue usage by angry negotiators under different levels of anger is unexplored in negotiation literature. In this paper, we address this gap by conducting a distributive electronic negotiation experiment and studying the usage of cues (statements and para-linguistic cues including emoticons) by angry negotiators while interacting with their counterpart (computer). We report that participants tend use more para-cues, especially emoticons, as their anger intensity increases and that emoticons have the ability to replace other para-cues while composing angry messages. The findings provide promising inputs on design of user interfaces for electronic negotiation systems.

Keywords: Emoticons · Emotions · Anger · Negotiations · Cues · Anger intensity

1 Introduction

Emotion in negotiation has attracted a lot of attention in the past ten years [1–3, 51] and among all the emotions, the role of anger in negotiation has received considerable focus [3–6]. Usage of emotions such as threats [7] and other extreme form or anger [8–10] and their impact in terms of eliciting concessions [3, 5, 11, 12] were explored in detail. While anger is recognized and analyzed to a great extent, not much attention was given to the study of negotiator behavior under varying anger levels (i.e. anger intensity). In this study, we address this gap by studying message communication behavior of angry negotiators during an electronic negotiation as they go through varying degrees of anger. With modern UIs and Web 2.0 technologies, the options to communicate emotion through the Instant Messaging (IM) systems has improved many fold. Pictures, emoticons, emojis and other pictorial representations are available to be used along with written communication that greatly improves the ability to convey affect in addition to the usage of other traditional cues. In spite of all this development, the question of how angry negotiators leverage emoticons and other para-linguistic cues to convey anger and anger intensities have remained largely unexplored. As modules containing emoticons and other para-linguistic cues are now routinely bundled with

© Springer Nature Switzerland AG 2019
D. C. Morais et al. (Eds.): GDN 2019, LNBIP 351, pp. 178–192, 2019.
https://doi.org/10.1007/978-3-030-21711-2_14

chat software in mobile phones and computers, a better understanding of its usage can be crucial input in designing user friendly interfaces for electronic negotiation and a valuable contribution to the existing literature. Hence, we propose a comprehensive study on how the different types of para-linguistic cues are used by angry negotiators to express different anger intensities in a synchronous electronic negotiation. In the process, we attempt to answer the following research questions:

Research Questions:
How are paralinguistic cues used in conveying anger?
What is the relationship between paralinguistic cues and anger intensity?

The usage of cues in Computer-mediated-communication (CMC) has attracted researcher's attention for some time resulting in several categorizations. As early as 1980, Carey categorized nonverbal cues in CMC into five types: vocal spelling, lexical surrogates, spatial arrays, manipulation of grammatical markers, and minus features [21]. Vocal spelling includes extended emphasis such as "weeeeelllllll" and "yessss" while lexical surrogates use non-standard spelling such as "mhmm" and "uh huh" to mimic vocal intonation or tone. Text based emoticons using keyboard characters such as :-) for smile and :-D for laughing were categorized as spatial arrays and are used to represent facial expressions. Methods to indicate pauses (...), attitude or surprise (!!!) and tone of voice (SHOUT) were categorized as Manipulated grammatical markers. Carey's categorization also included Minus features refer to omission of certain language standards that are commonly expected such as lack of capitalization at the beginning of a sentence. Usage of capital letters, asterisks, blank spaces, or character repetitions, as well as combinations of these devices were also reported as writing styles [22]. Other cues identified include para and prosodic cues such as asterisks, capitalized words, repeating letters [23] and italicized words [24]. Emoticons are a relatively recent addition to the cue toolbox. Emoticons are defined in several ways, as string of characters that convey a particular emotion when viewed sideway [25], as pictographs [26] and as a creative way to add expression to text-based communication [27]. Modern emoticons may be a static or animated and incorporate several common emotional states to enable precise communication of emotion.

In addition to studying the usage of para cues, researchers have also focused on their ability to convey affect. Analyzing instant messaging conversations, Hancock et al. found that exclamation marks were a significant predictor of whether the receiver believed that the sender is in a positive mood [28]. Emoticons were also found to be a key cue in interpreting sender's emotion [26, 29, 30]. Riordan et al. analyzed cue usage using LIWC [32] (Linguistic Inquiry and Word Count) in five corpora downloaded from the internet [31]. They found that cues were used predominantly in disambiguation of a message, regulation of an interaction, expressing effect and strengthening message content. The ability of para-cues in general, in communicating affect in CMC was investigated by Harris et al. [33]. A range of emotion words, linguistic markers and paralinguistic cues were investigated on their ability to convey emotion in emails, with the conclusion that the number of emotion cues used is directly proportional to the strength of the sender's emotion as perceived by the receivers. Trends in emoticon usage in short internet chats showed that they are mostly used to convey emotion, strengthen a message and to express humor. They are used more when participants are communicating with their friends as opposed to strangers and in more positive context than in negative ones. In short, emoticons are used in a way similar to facial behavior in Face-to-Face (F2F) communication with respect to social context and interaction partner [34]. They are increasingly recognized as a way to indicate writer's moods or feelings in CMCs [35].

Yet, in the domain of negotiation, an understanding of influence of para-cues is lagging. A recent work in this field involves a study on usage of emoticons in synchronous and asynchronous chat communication in an electronic negotiation setting [36]. This study found that emoticons support and supplement text messages, increases the communication of positive affect in asynchronous negotiations and decreases communication of negative affect in synchronous negotiations. Apart from this, we could not find any literature exploring the significance of para-cues in conveying effect in electronic negotiations. Even this study [36] was focused on the impact of only emoticons, ignoring a larger group of para-cues (e.g. vocal spelling, lexical surrogates, spatial arrays, capitalized words, manipulation of grammatical markers, minus features) and was not focused on anger intensity. While [33] considered a larger set of cues, it did not include emoticons (static and dynamic), was focused on emotion perception and not emotion expression, email based and not in the negotiation domain.

In our work, we incorporate a wider range cues including animated emoticons and focuses exclusively on anger intensity. This study also differs from [33] by including animated emoticons in addition to other cues, focuses on their impact on message composition behavior (instead of message perception), and takes place in an electronic negotiation environment (instead of email based general CMC).

2 Theoretical Model

In F2F communication, several cues such as facial expression, body posture and speech patterns can readily be used to convey the intended message of the speaker [13, 14]. In contrast to F2F, electronic communication is considered cold and anonymous [15] and absent of all these stimuli which aid in identification of emotion. Two opposing theories exist with regard to communication using electronic media: 'cues filtered-out' and 'cues filtered-in'. The 'cues filtered-out' theory [16] argues that in computer mediated communication (CMC) there is a reduction in social cues about the negotiating counterparts such as their experiences, situations, perceptions and context and are dependent solely on the information exchanged via the communication channel. Important non-verbal cues containing rich information are unable to be transferred across electronic media. For example, there is an increased incidence of flaming when using computer mediated communication as compared to F2F negotiation, and this difference is attributed to the inability to transfer social cues from one negotiator to another in an electronic negotiation setting [17, 18].

'Cues filtered-in' theory [19], on the other hand, states that rich affective information can be transferred using text-based messages [20]. Theories that oppose the 'cues-filtered-out' classification, such as Social Information Processing (SIP), state that with newer, multimedia forms of communication it is possible to achieve the same level of impressions of others and develop relationship as off-line communication [16] by using whatever cue is available in the chosen communication channel, but communicators would need more time to accomplish this objective. While there is some empirical evidence to support each viewpoint, a recent review concludes that expression of emotion is similar in both, off- and on-line modes, and found no indication that CMS is a less emotional medium compare to F2F [46].

While the above theories mostly deal with affect and relationships, it is still unclear how affect intensities are communicated in CMC in general, and in a negotiation context in particular. Media Richness Theory [47] defines richness of a medium by *four dimensions,* namely: (i) number of cue systems; (ii) immediacy of feedback; (iii) potential for natural language; and, (iv) message personalization. Incorporating multiple cue systems, synchronous sender-receiver exchanges, the ability to converse in natural languages, and the ability to personalize each message to the participant all contribute to a richer medium. According to this formulation, F2F communication is the richest medium followed by telephone, letters, and memos. In our experiment, we attempt to enhance the richness of each of the aforementioned four dimensions. Specifically, the number of cue systems is increased to include natural language statements, static and animated emoticons, other para-cues (e.g. vocal spelling, lexical surrogates, manipulated grammatical markers), and minus features. The experiment involves a synchronous negotiation exercise, making it a simultaneous bi-directional interaction (similar to F2F). Participants can choose multiple natural language statements to compose their messages and they interact directly with their counterparts. On the basis of this rich setup, in the next section we articulate various hypotheses to study the communication of anger intensity.

3   Hypothesis

In this paper, we study how an angry negotiator communicates emotion and emotion intensity using various para-cues. For hypothesis formulation, we combine the cues into three categories: text messages, emoticons (static and animated emoticons), and other para-cues (vocal spelling, lexical surrogates and manipulated grammatical markers). Usage of para-cues may convey tone to the message and facilitate the communication of type and degree of emotion [34]. In [37], the hypothesis that emoticons will supplement text messages in electronic negotiation was supported. In the study of corpus of CMCs [32], it was found that usage of para-cues such as capitalized words, repeated punctuations, emoticons and combined cues were common. Cues are used together frequently. Capitalized word was frequently used with repeating exclamation marks, asterisks, repeating question marks, repeating letters and emoticon. Three-way usage of capitalized words with combined question mark and exclamation point was also found to occur. Emoticons, capitalized words, asterisks, underscores, angled brackets, curly braces exclamation marks, repeating letters, repeating exclamation marks were also used with one another. However, how para-cues and emoticons are used together in expressing anger is not explored in detail. We hypothesize that, if participants choose to use emoticons, they will no longer feel a need to support it with other para-cues. Conversely, usage of other para-cues will not necessitate the usage of emoticons. Hence,

H1a: There will be a negative correlation between emoticons and other para-cues usage in angry messages.
H1b: There will be a negative correlation between static emoticons and other para-cues usage in angry messages.
H1c: There will be a negative correlation between animated emoticons and other para-cue usage in angry messages.

Previous research has shown that participants have also successfully detected emotion in CMC using the metadata of the messages such as message length, usage of negative terms and message exchange rate [37]. "Number of cues" was also used as a cue to study message communication. In [33], the number of para-cues contained in positive messages was found to be positively correlated with valence and degree of emotion by receivers. Due to the difficult context setup by a distributive negotiation, we expect the participants to use the number of cues to express various anger intensity levels. Specifically,

H2a: In angry messages, anger intensity will be positively related with the number of cues used.
H2b: In angry messages, anger intensity will be positively related with the number of emoticons used.
H2c: In angry messages, anger intensity will be positively related with the number of other para-cues (excluding emoticons) used.

4 Experiment

We took as basis the multi-round electronic negotiation used by Van Kleef [3] and made modifications to suit our needs. The object of the negotiation is a used cell-phone. Issues under consideration were price, warranty and service (Table 1). Participants were informed that they will be randomly assigned the role of a buyer or seller but, in reality, all of them were assigned the role of a buyer. In this aspect, we deviate from the setup used in [3]. With e-commerce becoming common, almost everyone would have assumed the role of a buyer at some point of time making it easier to relate to the task. Research indicates that such role-reversals do not impact concession making [38].

Table 1. Participant's issue options

Price ($)	Warranty (months)	Service (months)
150	1	1
145	2	2
140	3	3
135	4	4
130	5	5
125	6	6
120	7	7
115	8	8
110	9	9

Personalized utility: While payoffs are assigned to each issue choice in [3], we use a utility function based on user's preferences to calculate individual payoff, which enhances the relevance of the offer. Participant's preference for price, warranty and service were captured by asking to rate them individually on a scale of 0 to 1 and Eq. 1 is used to calculate individual utility. The utility ranges from 0 to 1, with a utility of 1 being the best utility and 0 being the worst utility for the buyer.

$$u_{i,r} = \frac{p_{pref,i} * (P_{max} - p_r)}{P_{max} - P_{min}} + \frac{w_{pref,i} * (w_r - W_{min})}{W_{max} - W_{min}} + \frac{s_{pref,i} * (s_r - S_{min})}{S_{max} - S_{min}} \quad (1)$$

Where,

$u_{i,r}$ is the utility of user i at round r,

$p_{pref,i}, w_{pref,i}, s_{pref,i}$ is the preference of price, warranty and service respectively of user i,

$P_{max}, W_{max}, S_{max}$, are the maximum price, warranty and service that a user can select,

$P_{min}, W_{min}, S_{min}$, are the minimum price, warranty and service that a user can select,

p_r, w_r, s_r is the price, warranty and service selected by a user in round r,

$p_{pref,i} + w_{pref,i} + s_{pref,i} = 1$,

$0 \leq p_{pref,i}, w_{pref,i}, s_{pref,i} \leq 1$.

4.1 Experimental Steps

A web-based application that takes the user through a series of steps through a wizard was developed. The different steps are explained below:

Step 1 (Demographic data): Gender, age group, country and state of birth, country and state of residence and job level of the participants were captured.

Step 2 (Summary of steps): Summary of the instructions with a flowchart of the negotiation process was shown in step 2.

Step 3 (Inter-issue preferences): Preferences for price, warranty and service were captured.

Step 4 (Role assignment): The participants were asked for wait while the computer supposedly assigns them to buyer and seller roles and pairs them up.

Step 5 (Offer generation from the e-negotiation system): This is the main negotiation web-page, and its layout is shown in Fig. 1. The seller (computer) provided the first offer. The offer was shown in a text box as "My first offer is: price = Rs. 150, warranty = 1 month and service = 1 month". Typographical errors and more subtle errors were introduced to make the responses more human. The utility of the seller's offer to the buyer and buyer's previous offer and its corresponding utility was also displayed for easy comparison and decision making. The buyer is asked whether he accepts the offer, which he can accept or reject through a drop-down box.

Initially, on page load, only the seller's offer, buyer's previous offer, their utilities and the question of whether the participants accept or reject the offer is displayed to avoid confusion. If the offer is rejected, the issue options table, utilities of counter-offer, emotion and emotion intensity elicitation questions and a chat-box along with predefined statements and emoticons are displayed. Buyer information is shown in blue and seller information in red. The participant selects the counteroffer from the HTML table. The corresponding utility, is calculated in real time and displayed for comparison. The buyer is prevented from submitting an offer that has lower utility than the current computer's offer.

Text boxes were provided to the participant to record their responses while submitting the counter offers. The cues featured three angry, two happy and one neutral sentence. Top three angry sentences and top two happy sentences were selected based on their perceived intensity from a previous study [39]. Further a palette paralinguistic cues were also selected. One happy and one angry static emoticon, one happy and one angry animated emoticon were selected from the top recommendations from a Google search. Manipulated grammatical markers (e.g. 'caps lock' and '!!'), vocal spelling manipulations (e.g. goood offer, baaad offer), lexical surrogates (e.g. ugghh) were selected as para-cues based on literature [21]. The participants were mandated to choose at least one text statement and at least one cue from the list. However, a full-stop cue (i.e. '.') was also included in the pallet of para-cues, in case the user did not want to select any cue. The position of the statements and cues were randomized for each round. Along with the statement, the emotion and emotion intensity of the participants were also captured. Once the participant provides all the information, we move on to Step 6.

If the computer's offer is accepted, the participants were asked to provide only their emotion, emotion intensity and message to the counterpart. Then the participant is directed to an animated screen displaying the message, "Please wait while your opponent evaluates your offer and responds". After 1.5 min, a message "Thank you for accepting the offer", is displayed and the negotiation ends. The layout of the page and the descriptions of the corresponding sections are provided in Fig. 1.

Step 6 (Counteroffer submission): Once the participants submit the counteroffer, an animated screen displaying "Please wait while your opponent evaluates your offer and responds" is displayed. After 1.5 min, the negotiation page (see Fig. 1) re-appears for the next round with the new counter-offer. The negotiation ends if the buyer accepts the offer or if the utility of the counter offer provided by the buyer equals the computer's next offer or if six rounds are completed.

4.2   Anger Induction Strategy

Anger is induced by incorporating long wait time between rounds and a dynamic distributive strategy. Participants in distressed situation may interpret a long response time by the counterpart as a personal attack [8, 40] and has been known to induce anger [9]. Hence a 1.5 min time gap was introduced. Further, a dynamic distributive strategy, unique to this experiment, based on the participant's preferences was used to induce anger. Prior to the negotiation, the participants were asked to provide the importance of each of issue in a scale of 0 to 1 and they are ranked accordingly. The participants were told that sellers always make the first offer. And as the buyer role is always assigned to the participants. the computer makes the first offer in all cases. Distributive negotiations are characterized by extreme first offers, no or small number of concessions, going back on offers and lesser amount of concessions as compared to integrative negotiations. This characterization is implemented in different rounds as follows,

Round 1: start with the extreme offer ($150 for price, 1 month for warranty and 1 month of service). This is the worst deal to the participant with a utility of zero (as per Eq. 1). If the offer is rejected, the participant is asked to choose a counteroffer and move to round 2.

Round 2: The revised offer by the computer keeps the same value for two issues and concedes one unit on the third issue. The concession is made on the buyer's least preferred issue. If this offer is rejected by the buyer, he/she will be asked to provide a counteroffer to move on to round 3.

Round 3: The computer goes back to the offer provided at round 1, negating the concession offered in round 2. A rejection of this offer takes the participant to round 4.

Round 4: The computer repeats the offer given in round 2. Rejection of this offer lead to round 5.

Round 5: No change in the computer's offer.

Round 6: A small concession of one unit on the participants second least preferred issue is added to the round 4 offer. Table 2 lists an implementation where the buyer's issue preference is price, warranty and service in that order.

Table 2. Strategy implementation for buyer preference of price > warranty > service

Round 1: Price = $150, Warranty = 1 month, Service = 1 month.
Round 2: Price = $150, Warranty = 1 month, Service = 2 months.
Round 3: Price = $150, Warranty = 1 month, Service = 1 months.
Round 4: Price = $150, Warranty = 1 month, Service = 2 months.
Round 5: Price = $150, Warranty = 1 month, Service = 2 months.
Round 6: Price = $150, Warranty = 2 months, Service = 2 months.

The experiment was conducted in the classroom and over phone. A presentation was prepared and explained to the participants in-person (in the case of classroom experiment) and over phone (in case of remote experiment). They were told that the intent is to study negotiation where the participants do not see each other, that they may be a buyer or seller and the task is to negotiate the sale of a cell phone. They were informed that the negotiation will end if an agreement is reached or when time runs out. No mention was made on the time limit. Information on how to select preferences and the meaning of utility were explained. Screenshots of the web pages were included in the presentation to familiarize the participants. The participants were then led to a lab for the experiment. The computers in the lab were spaced sufficiently and a proctor ensured that the participants do not speak or interact in any other way.

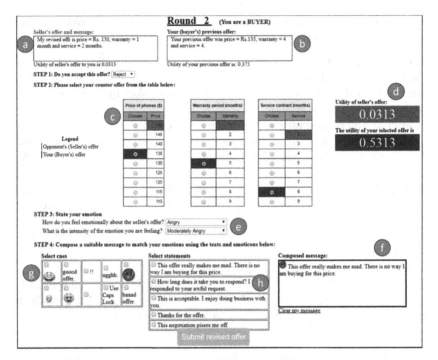

Fig. 1. Layout of the main negotiation page. *Sec. a: Seller offer communication; b: Previous offer display for comparison; c: Options table with the seller offer coded in red and buyer offer in blue; d: Realtime display of buyer and seller utility as per user selection of price, warranty and service e: Emotion and emotion intensity capture. "Angry", "Happy", "Sad" and "Other" are the emotion choices; f: Chat-box for viewing composed message. g: Para-linguistic cue display for message composition; h: Angry, happy and neutral statements for message composition;* (Color figure online)

5   Results

Ninety-six participants took the experiment, out of which 66.67% were male and 33.33% were female. Angry emotion was coded as 1 and non-angry emotion was coded as 0 and a logistic regression was carried out with emotion as dependent variable and round as independent variable. The results gave a positive effect of round id on emotion with each increase in round resulting in a 33% likelihood of participants getting angry suggesting that the experimental manipulation was successful in inducing anger. A total of 271 angry statements were recorded out of 390 statements (70%), providing more evidence that the anger induction was successful. The distribution of cues among the corpus of angry statements is shown in Fig. 2.

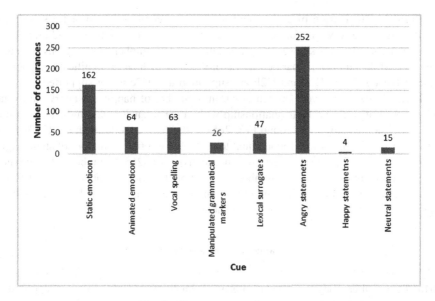

Fig. 2. Cue occurrence in corpus

There was a significant negative correlation between emoticons and other para-cues ($r = -0.66$, $p < .001$, $n = 271$). Further, a significant negative correlation was found between static emoticons and other para-cues ($r = -0.47$, $p < .001$, $n = 271$) and between animated emoticons and other para-cues ($r = -0.20$, $p < .001$, $n = 271$) providing support for H1a, H1b and H1c. Further decomposing emoticons into angry and happy static and animated emoticons found that angry static emoticons were negatively correlated with other para-cues (-0.45, $p < 0.001$, $n = 271$), angry animated emoticons were negatively correlated with other para-cues (-0.21, $p < 0.001$, $n = 271$). Happy emoticons do not have any significant correlation with other para-cues (Table 3).

Table 3. Relationship between anger intensity and number of cues

Dependent variable: Anger intensity				
Intercept	Independent variable: Number of			R^2 (Adj.)
	Cues	Emoticons	Other para-cues	
3.0416[***] (0.1785)	0.3280[**] (0.1167)	–	–	0.0249** $F(1, 269) = 8.8960$
2.8201[***] (0.1714)	–	0.8569[***] (0.1394)	−0.1095 (0.1307)	0.1417[***] $F(2, 268) = 23.3$

Note: Two equations modeled are:
Anger intensity = intercept + number of cues
Anger intensity = intercept + number of emoticons + number of other para-cues
[*]$p < 0.05$, [**]$p < 0.01$, [***]$p < 0.001$. Standard error in parenthesis.

Results of regression of anger intensity on number of cues returned a significant result. Anger intensity was further regressed onto number of emoticons and number of other para-cues. Results showed a significant impact of number of emoticons on anger intensity while number of para-cues was not found to have a significant impact on anger intensity. Thus, H2a and H2b are supported and H2c is not supported.

Further, emoticons were broken down into number of happy and angry static and animated emoticons and their relationship with anger intensity was studied. A significant regression equation was found $F(5, 265) = 10.76$, $p < 0.000$, with an R^2 (Adj.) of 0.1531. Anger intensity was found equal to $2.7582 + (1.0646 *$ number of static angry emoticons$) - (0.4248 *$ static happy emoticons$) + (0.6255 *$ number of animated angry emoticons$) + (0.4414 *$ number of animated happy emoticons$) - (0.0889 *$ number of other para-cues$)$. The number of angry static and animated emoticons were significant predictors of anger intensity (Table 4).

Table 4. Results summary

Hypothesis	Results
H1a: There will be a negative correlation between emoticons and other para-cues usage in angry messages	Supported
H1b: There will be a negative correlation between static emoticons and other para-cues usage in angry messages	Supported
H1c: There will be a negative correlation between animated emoticons and other para-cue usage in angry messages	Supported
H2a: In angry messages, anger intensity will be positively related with the number of cues used	Supported
H2b: In angry messages, anger intensity will be positively related with the number of emoticons used	Supported
H2c: In angry messages, anger intensity will be positively related with the number of other para-cues (excluding emoticons) used	Not supported

6   Discussion

Angry statements and static emoticons together account for 67% of the cues. Only three instances of happy static emoticons, two instances of happy animated emoticons and four instances of happy statement usage were found, suggesting reduced composition of ironic or sarcastic messages and that participants leave little room for ambiguity in expressing anger. While [36] found that emoticons act as supplement to text messages, our result show that emoticons can replace other para-cues as well. The empherical evidence suggests that angry emoticons and para-cues are used in a mutually exclusive manner, thereby suggesting that both static and animated angry emoticons have the ability to communicate anger as well as the other para-cues. [33] reported that message receivers recorded a higher degree of sender's emotion with the increase in the number of emotion cues. Our regression result between anger intensity and number of cues adds to this result by suggesting that message composers indeed use the number of cues itself as a cue to convey higher emotion (anger) intensity. Emoticons score over other para-cues in this respect with the usage of more number of angry emoticons signifying higher anger intensities.

Our results point to two findings. First, both static and animated angry emoticons have the ability to replace other para-cues while communicating anger. Second, both the number of static and animated angry emoticons can be used as predictors of anger levels while para-cues do not have any role to play. While researchers have studied static emoticons before, our study find that animated emoticons are also useful in composing angry messages and the also have value in predicting anger levels. The Emotion as Social Information (EASI) [48, 49] model argues that participants deduce social information of their counterpart through emotions and they use such information to guide their actions. As emotion intensity (anger intensity, in our case) is a key component of emotion, it is possible that intensity levels are also processed by the recipients and used to decide their responses. Our findings suggest that, changes in the number of angry emoticons in a corpus of received messages can be used to determine shifts in anger levels of counterparts and participants need not pay much attention to other para-cues. As anger can lead to impasse or breakdown of negotiation and have adverse impact on relationships, any tool or feature that mitigates these negative effects and clarify emotional states adds value to negotiation research. Designers of negotiation systems supporting distributive negotiations may benefit by focusing more on angry static and animated emoticons instead of other para-cues. Features that automatically convert ASCII based angry emoticons to pictorial ones (as is already available in several software such as MS Word) may enable participants to convey different anger intensity levels clearly and unambiguously and may have the potential to lead to better negotiation outcomes.

7 Conclusion

Anger plays a significant role in negotiations. In this paper we addressed a dimension of anger, anger intensity, by investigating message composition behavior of angry negotiators under varying levels of anger. Our results show both static and animated angry emoticons play a central role in the communication of anger levels. Communication systems designed for distributive negotiations need to consider both these forms of emoticons as a key cue in the expression of anger intensity.

Our study is not without limitations. The personalized utility function is one way of capturing the utility of the participants. Adopting other approaches might result in different conclusions. While a presentation of the preference elicitation screen and an explanation was given to the participant before the experiment to enhance their understanding, there is a chance that other factors might also influence their understanding of the weights. A corpus of 391 statements might be too low to arrive at definitive conclusions. While care was taken in the experiments to hide the real intent, the fact that emotion and emotion intensity information is elicited in all rounds might have given away the goal of the experiment and this might have had an impact on the result. Emotion and emotion intensity are elicited as self-reports. Other forms of elicitation such as facial electromyography and skin conductance responses [50] might provide a more accurate measure of emotion and intensity. While looking at only one emotion, we indirectly subscribed to the Discrete Emotion Theory [41, 42]. However, it is entirely possible that participants may be experiencing a range of emotions, as proposed by the Dimensional model of emotion [43–45].

References

1. Sinaceur, M., Kopelman, S., Vasiljevic, D., Haag, C.: Weep and get more: when and why sadness expression is effective in negotiations. J. Appl. Psychol. **100**(6), 1847–1871 (2015)
2. Van Kleef, G.A., De Dreu, C.K.W., Manstead, A.S.R.: Supplication and appeasement in conflict and negotiation: the interpersonal effects of disappointment, worry, guilt, and regret. J. Pers. Soc. Psychol. **91**(1), 124–142 (2006)
3. Van Kleef, G.A., De Dreu, C.K., Manstead, A.S.: The interpersonal effects of anger and happiness in negotiations. J. Pers. Soc. Psychol. **86**(1), 57–76 (2004)
4. Van Kleef, G.A., De Dreu, C.K.W., Pietroni, D., Manstead, A.S.R.: Power and emotion in negotiation: power moderates the interpersonal effects of anger and happiness on concession making. Eur. J. Soc. Psychol. **36**(4), 557–581 (2006)
5. Steinel, W., Van Kleef, G.A., Harinck, F.: Are you talking to me?! Separating the people from the problem when expressing emotions in negotiation. J. Exp. Soc. Psychol. **44**(2), 362–369 (2008)
6. Adam, H., Shirako, A., Maddux, W.W.: Cultural variance in the interpersonal effects of anger in negotiations. Psychol. Sci. **21**(6), 882–889 (2010)
7. Sinaceur, M., Van Kleef, G.A., Neale, M.A., Adam, H., Haag, C.: Hot or cold: is communicating anger or threats more effective in negotiation? J. Appl. Psychol. **96**(5), 1018–1032 (2011)
8. Johnson, N.A., Cooper, R.B., Chin, W.W.: Anger and flaming in computer-mediated negotiation among strangers. Decis. Support Syst. **46**(3), 660–672 (2009)
9. Reinig, B.A., Briggs, R.O., Nunamaker, J.F.: Flaming in the electronic classroom. J. Manag. Inf. Syst. **14**(3), 45–59 (1998)
10. Johnson, N.A., Cooper, R.B., Chin, W.W.: The effect of flaming on computer-mediated negotiations. Eur. J. Inf. Syst. **17**(4), 417–434 (2008)
11. Sinaceur, M., Tiedens, L.Z.: Get mad and get more than even: when and why anger expression is effective in negotiations. J. Exp. Soc. Psychol. **42**(3), 314–322 (2006)
12. Van Kleef, G.A., De Dreu, C.K., Manstead, A.S.: The interpersonal effects of emotions in negotiations: a motivated information processing approach. J. Pers. Soc. Psychol. **87**(4), 510–528 (2004)
13. Scherer, K.R.: Vocal communication of emotion: a review of research paradigms. Speech Commun. **40**(1), 227–256 (2003)
14. Coulson, M.: Attributing emotion to static body postures: recognition accuracy, con-fusions, and viewpoint dependence. J. Nonverbal Behav. **28**(2), 117–139 (2004)
15. Danet, B., Ruedenberg-Wright, L., Rosenbaum-Tamari, Y.: Hmmm... where's that smoke coming from? J. Comput.-Mediated Commun. **2**(4), JCMC246 (1997)
16. Walther, J.B., Parks, M.R.: Cues filtered out, cues filtered. In: Handbook of Interpersonal Communication, pp. 529–563 (2002)
17. Hiltz, S.R., Turoff, M., Johnson, K.: Experiments in group decision making, 3: disinhibition, deindividuation, and group process in pen name and real name computer conferences. Decis. Support Syst. **5**(2), 217–232 (1989)
18. Sproull, L., Kiesler, S.: Reducing social context cues: electronic mail in organizational communication. Manag. Sci. **32**(11), 1492–1512 (1986)

19. Walther, J.B., Loh, T., Granka, L.: Let me count the ways. J. Lang. Soc. Psychol. **24**(1), 36–65 (2005)
20. Moore, D.A., Kurtzberg, T.R., Thompson, L.L., Morris, M.W.: Long and short routes to success in electronically mediated negotiations: group affiliations and good vibrations. Organ. Behav. Hum. Decis. Process. **77**(1), 22–43 (1999)
21. Carey, J.: Paralanguage in computer mediated communication. In: Proceedings of the 18th Annual Meeting on Association for Computational Linguistics, pp. 67–69. (1980)
22. Spitzer, M.: Writing style in computer conferences. IEEE Trans. Prof. Commun. **29**(1), 19–22 (1986)
23. Crystal, D.: Language and the Internet. Cambridge University Press, Port Chester (2001)
24. Fox, A.B., Bukatko, D., Hallahan, M., Crawford, M.: The medium makes a difference: gender similarities and differences in instant messaging. J. Lang. Soc. Psychol. **26**(4), 389–397 (2007)
25. Danesi, M.: Dictionary of Media and Communication. Routledge, New York (2009)
26. Thompson, P.A., Foulger, D.A.: Effects of pictographs and quoting on flaming in electronic mail. Comput. Hum. Behav. **12**(2), 225–243 (1996)
27. Luor, T.T., Wu, L.L., Lu, H.P., Tao, Y.H.: The effect of emoticons in simplex and complex task-oriented communication: an empirical study of instant messaging. Comput. Hum. Behav. **26**(5), 889–895 (2010)
28. Hancock, J.T., Landrigan, C., Silver, C.: Expressing emotion in text-based communication. In: Proceedings of the ACM Conference on Human Factors in Computing Systems (CHI 2007), pp. 929–932 (2007)
29. Byron, K., Baldridge, D.: Toward a model of nonverbal cues and emotion in email. In: Academy of Management Proceedings, vol. 1 (2005)
30. Utz, S.: Social information processing in MUDs: the development of friendships in virtual worlds. J. Online Behav. **1**(1) (2000)
31. Riordan, M.A., Kreuz, R.J.: Cues in computer-mediated communication: a corpus analysis. Comput. Hum. Behav. **26**(6), 1806–1817 (2010)
32. Pennebaker, J.W., Chung, C.K., Ireland, M., Gonzales, A., Booth, R.J.: The development and psychometric properties of LIWC2007. LIWC.net, Austin (2007)
33. Harris, R.B., Paradice, D.: An investigation of the computer-mediated communication of emotions. J. Appl. Sci. Res. **3**(12), 2081–2090 (2007)
34. Derks, D., Bos, A.E., Von Grumbkow, J.: Emoticons in computer-mediated communication: social motives and social context. CyberPsychol. Behav. **11**(1), 99–101 (2008)
35. Gajadhar, J., Green, J.: The importance of nonverbal elements in online chat. Educause Q. **28**(4), 63–64 (2005)
36. Gettinger, J., Koeszegi, S.T.: More than words: the effect of emoticons in electronic negotiations. In: Kamiński, B., Kersten, G.E., Szapiro, T. (eds.) GDN 2015. LNBIP, vol. 218, pp. 289–305. Springer, Cham (2015). https://doi.org/10.1007/978-3-319-19515-5_23
37. Hancock, J.T., Gee, K., Ciaccio, K., Lin, J.M.H.: I'm sad you're sad: emotional contagion in CMC. In: Proceedings of the 2008 ACM Conference on Computer Supported Cooperative Work (CSCW 2008), pp. 295–298. ACM, New York (2008)
38. Bagchi, R., Koukova, N.T., Gurnani, H., Nagarajan, M., Oza, S.S.: Walking in my shoes: how expectations of role reversal in future negotiations affect present behaviors. J. Mark. Res. **53**(3), 381–395 (2016)
39. Venkiteswaran, S., Sundarraj, R.P.: Perceived anger intensity in electronic negotiation. In: Proceedings of the 17th International Conference on Group Decision and Negotiation (2016)
40. Beck, A.T., Deffenbacher, L.J.: Prisoners of hate: the cognitive basis of anger, hostility and violence. J. Cogn. Psychother. **14**(2), 201–203 (2000)
41. Ekman, P.: An argument for basic emotions. Cogn. Emot. **6**(3-4), 169–200 (1992)

42. Colombetti, G.: From affect programs to dynamical discrete emotions. Philos. Psychol. **22** (4), 407–425 (2009)
43. Posner, J., Russell, J.A., Peterson, B.S.: The circumplex model of affect: an integrative approach to affective neuroscience, cognitive development, and psychopathology. Dev. Psychopathol. **17**(3), 715–734 (2005)
44. Rubin, D.C., Talarico, J.M.: A comparison of dimensional models of emotion: evidence from emotions, prototypical events, autobiographical memories, and words. Memory **17**(8), 802–808 (2009)
45. Watson, D., Tellegen, A.: Toward a consensual structure of mood. Psychol. Bull. **98**(2), 219–235 (1985)
46. Derks, D., Fischer, A.H., Bos, A.E.R.: The role of emotion in computer-mediated communication: a review. Comput. Hum. Behav. **24**(3), 766–785 (2008)
47. Daft, R.L., Lengel, R.H.: Organizational information requirements, media richness and structural design. Manag. Sci. **32**(5), 554–571 (1986)
48. Van Kleef, G., De Dreu, C., Manstead, A.: An interpersonal approach to emotion in social decision making: the emotions as social information model. Adv. Exp. Soc. Psychol. **42**, 45–96 (2010)
49. Van Kleef, G.A.: How emotions regulate social life: the emotions as social information (EASI) model. Curr. Dir. Psychol. Sci. **18**(3), 184–188 (2009)
50. Leppänen, I., Hämäläinen, R.P., Saarinen, E., Viinikainen, M.: Intrapersonal emotional responses to the inquiry and advocacy modes of interaction: a psychophysiological study. Group Decis. Negot. **27**(6), 933–948 (2018)
51. Martinovsky, B. (ed.): Emotion in Group Decision and Negotiation, vol. 7. Springer, Dordrecht (2015). https://doi.org/10.1007/978-94-017-9963-8

Opinion Dynamics Theory Considering Trust and Suspicion in Human Relations

Akira Ishii[1,2(✉)] [ID]

[1] Tottori University, Koyama, Tottori 680-8552, Japan
ishii.akira.t@gmail.com
[2] Center for Computational Social Science, Kobe University,
2-1 Rokkodai, Nada, Kobe 657-8501, Japan

Abstract. We present a new opinion dynamics theory including trust and distrust among people of society, and consideration of mass media effects. The opinion axis is assumed to be one dimension. Mutual trust and distrust relations is assumed to be asymmetric. This theory is a general theory that can describe society that can deal not only with consensus building but also division of society.

Keywords: Conflict · Distrust · Simulation

1 Introduction

Opinion dynamics is a theory to analyze how many human opinions converge by exchange of opinions. It has been studied in sociology etc. since ancient times [1–6,19]. The theory of opinion dynamics is summarized by the comprehensive report of ref. [7,8], for example. The well-known representative ones are the voter model [9,10], the Ising model-like theory by Galam [11–13], the local majority decision theory by the same Galam [14,15], and the Bounded Confidence Model [16–18]. In the voter model, the Ising model and the local majority decision theory, each person's opinion is binary opinion, 0 or 1, or −1 to 1, whereas the Bounded Confidence model says that opinions are continuously distributed. In the Bounded Confidence Model, the conversion of opinions of people in society is implicitly expected.

Although this research is based on the Bounded Confidence Model as the Hegselmann-Krause theory [16], it extends the relationship of individual people to have both trusts and distrusts. Furthermore new opinion dynamics theory includes effects by mass media etc. Though the effects of distrust among people has been already suggested by Abelson-Bernstein [19] in 1963, such suggestion is not included in Hagselmann-Krause theory [16]. The early stage of this new theory has been appeared in ref. [20], the theory of the present paper is improved for the additional function of the strength of the coefficient of the relationship of individual people.

In this study, we propose a theory that expresses opinions as continuous values and deals with changes in the opinion values due to the exchange of

© Springer Nature Switzerland AG 2019
D. C. Morais et al. (Eds.): GDN 2019, LNBIP 351, pp. 193–204, 2019.
https://doi.org/10.1007/978-3-030-21711-2_15

opinions with others as assumed in the Bounded Confidence Model. Moreover, we assume that the opinion of each people can be both positive or negative values in contrast to the original Bounded Confidence Model. For example, in a study of Tweet on political situation in the United States, there is a study to classify political opinions from conservative to liberal by one-dimensional axis [22]. In this case, we can assume that the conservative opinion is positive/negative value and the liberal opinion is negative/positive value, for example.

In this research, we assume that differences in opinion can be represented by one-dimensional axis values as in this reference. Based on this theory, it is possible to express the division of opinion in society, assuming that opinions of people who disagree with each other are exchanged, and the opinions of both are further divided. Such a division of opinions is a phenomenon often seen on social media such as Twitter.

2   Modeling Opinion Dynamics

Our model is based on the original bounded confidence model of Hegselmann-Krause [16]. For a fixed agent, say i, where $1 \leq i \leq N$, we denote the agent's opinion at time t by $I_i(t)$. As shown in Fig. 1, person i can be affected by surrounding people. According to Hegselmann-Krause [16], opinion formation of agent i can be described as follows.

$$I_i(t+1) = \sum_{j=1}^{N} D_{ij} I_j(t) \tag{1}$$

This can be written in the following form.

$$\Delta I_i(t) = \sum_{j=1}^{N} D_{ij} I_j(t) \Delta t \tag{2}$$

where it is assumed that $D_{ij} \geq 0$ for all i, j in the model of Hegselmann-Krause. Based on this definition, $D_{ij} = 0$ means that the opinion of agent i is not affected by the opinion of agent j. In this theory, it is expected implicitly that the final goal of the negotiation among people is the formation of consensus.

However, in the real society in the world, the formation of consensus among people is sometimes very difficult. We can find many such examples in the international politics in the world history. Even in domestic problems, the opinions between people pursuing economic development and people claiming nature conservation are not compatible and agreement is difficult. Since it is not possible to define the payoff matrix for such serious political conflict, application of game theory may be difficult. Thus, in order to deal with problems that are difficult to form consensus among these people, it is necessary to include the lack of trust between people in our opinion dynamics theory. Here, as a result of exchanging opinions, consider the possibility that the opinions of two people with different opinions change move in different directions. We consider the distribution of

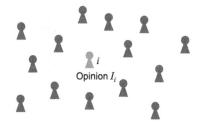

Fig. 1. Schematic illustration of opinion dynamics

opinions with in the positive and negative directions of a one-dimensional axis. In this case, the value range of $I_i(t)$ is $-\infty \leq I_i(t) \leq +\infty$. Here, we assume that $I_i(t) > 0$ means positive opinion and $I_i(t) < 0$ means negative opinion. In the limitation of Hegselmann-Krause model, one can assume that $1/2 \leq I_i(t) \leq 1$ corresponds to positive opinion and $0 \leq I_i(t) \leq 1/2$ corresponds to negative opinion. However, our definition is intuitive, easy to understand and easy to apply to various examples.

We modify the meaning of the coefficient D_{ij} as the coefficient of trust. We assume here that $D_{ij} > 0$ if there is a trust relationship between the two persons, and $D_{ij} < 0$ if there is distrust relationship between the two persons.

In contrast to our previous theory [20], we consider here that people disregard the opinion far removed from their opinions without agreeing or repelling. Also, opinions that are very close to himself/herself will not be particularly affected. To include the two effects, we use the following function instead of $D_{ij}I_j(t)$ as follows,

$$D_{ij}\Phi(I_i, I_j)(I_j(t) - I_i(t)) \tag{3}$$

where

$$\Phi(I_i, I_j) = \frac{1}{1 + exp(\beta(|I_i - I_j| - b))} \tag{4}$$

This function is Sigmoid function and it works as a smooth cut-off function at $|I_i - I_j| = b$. The typical graph of this function is shown in Fig. 2. Using this Sigmoid function, we assume that if the opinions of the two are too far apart, they will not be totally influenced by each other's opinion. Moreover, because of the factor $I_j(t) - I_i(t)$, the opinion $I_i(t)$ is not affected by the opinion $I_j(t)$ if the opinion $I_j(t)$ is almost same as the opinion $I_i(t)$.

For the factor $I_j(t) - I_i(t)$, we consider that $I_j(t) - I_i(t)$ gives the same effect whether $I_i(t)$ and $I_j(t)$ are both positive or both negative or either positive or negative. This is very natural that, for example, even between conservatives, there are intense debates between those with moderate conservatives and those with radical conservatives.

Influences of mass media and government statements can not be ignored in the formation of public opinion. Such mass media effect can also work even for

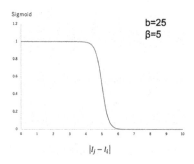

Fig. 2. The typical graph of the Sigmoid function (4) as smooth cut-off function.

negotiations of small size group. Since formula of our theory above is similar to the model of hit phenomena [21] where the popularity of certain topic is analyzed using the sociophysics model, we introduce here the effects of mass media similar to the way of ref. [21]. Let $A(t)$ be the pressure at time t from the outside and denote the reaction difference for each agent is denoted by the coefficient c_i. The coefficient c_i can have different values for each person and c_i can be positive or negative. If the coefficient c_i is positive, the person i moves the opinion toward the direction of the mass media. On the contrary, if the coefficient c_i is negative, the opinion of the person change against the mass media direction.

Therefore, including such mass media effects, the change in opinion of the agent can be expressed as follows.

$$\Delta I_i(t) = c_i A(t) \Delta t + \sum_{j=1}^{N} D_{ij} \Phi(I_i(t), I_j(t))(I_j(t) - I_i(t)) \Delta t \qquad (5)$$

We assume here that D_{ij} is an asymmetric matrix; D_{ij} and D_{ji}, $D_{ij} \neq D_{ji}$ and D_{ij} and D_{ji} can have different signs.

Long-term behavior requires attenuation, which means that topics will be forgotten over time. Here we introduce exponential attenuation. The expression is as follows.

$$\Delta I_i(t) = -\alpha I_i(t) \Delta t + c_i A(t) \Delta t + \sum_{j=1}^{N} D_{ij} \Phi(I_i(t), I_j(t))(I_j(t) - I_i(t)) \Delta t \quad (6)$$

3   Model Calculation

3.1   Opinion Dynamics for Two Agents

Let us first consider the case where the opinions of the two agents are the same. In the calculation below, we set $A(t)$ to be 0.005 as a constant value and the coefficient c_i is set to be unity. All calculations in this paper, we assume that $A(t)$ is constant for simplicity in order to pay attention to the effect of D_{ij}. In the actual simulation of the real society behaviors, the time-dependence of the external effect $A(t)$ is also significant.

First, we consider the case of two person, N = 2. Using our former model [20], the equations of the model two persons are as follows.

$$\Delta I_A(t) = c_A A(t)\Delta t + D_{AB} I_B(t)\Delta t \tag{7}$$

$$\Delta I_B(t) = c_B A(t)\Delta t + D_{BA} I_A(t)\Delta t \tag{8}$$

Using the model of the present paper, we can write down the equations for the case of N = 2 in the following way,

$$\Delta I_A(t) = c_A A(t)\Delta t + D_{AB}\Phi(I_A(t), I_B(t))(I_B(t) - I_A(t))\Delta t \tag{9}$$

$$\Delta I_B(t) = c_B A(t)\Delta t + D_{BA}\Phi(I_B(t), I_A(t))(I_A(t) - I_B(t))\Delta t \tag{10}$$

We name here the two agents A and B. In the first case, we set both D_{AB} and D_{BA} are positive. It means that the two person A and B trust each other. Thus, the opinion $I_i(t)$ moves in the positive direction as shown in Fig. 3. This means that by having a conversation with an agent of the same positive opinion, the two agents A and B will change its opinion to be more and more positive. Similarly, if the opinions of both agents are the same negative opinion, the opinions become more and more negative.

However, it is strange that people's opinions diverge to infinity over time. Therefore, this model can not be applied for cases of long time behavior.

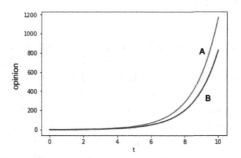

Fig. 3. Calculation result for N = 2. Adv = 0.5, $D_{AB} = 1.0$, $D_{BA} = 0.5$. The initial value is $I_A(0) = 0.005$, $I_B(0) = 0.2$.

In the revised model proposed in this paper, people is not influenced each other if their opinion is greatly distant. Also, it is not affected by opinions similar to his/her own opinion. Furthermore, it is assumed that the opinion will be influenced in proportion to the difference in strength of opinion between the two, $I_j(t) - I_i(t)$. Then, the calculation is corrected as shown in Fig. 4. As can be seen in Fig. 4, in this theory, the two opinions converge on one opinion, even if they have similar positive opinions.

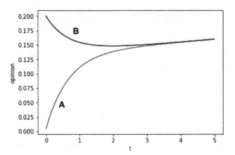

Fig. 4. Calculation result for N = 2. Adv = 0.005, $D_{AB} = 1.0$, $D_{BA} = 0.5$. The initial value is $I_A(0) = 0.005$, $I_B(0) = 0.2$.

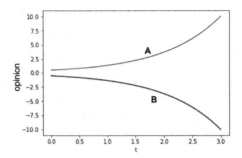

Fig. 5. Calculation result for N = 2. Adv = 0.5, $D_{AB} = 0.01$, $D_{BA} = 0.4$. The initial value is $I_A(0) = 0.005$, $I_B(0) = 0.2$.

Next, we consider the case where the opinions of the two agents are opposite: $I_A(t) > 0$ and $I_B(t) < 0$ where $D_{AB} < 0$ and $D_{BA} < 0$. The calculated result using the former model [20] is shown in Fig. 5. In this case, $D_{AB}I_B(t)$ is positive. Thus, the opinion of agent A changes to more a positive one. It means that, in a discussion with an agent in disagreement, as there is no trust relationship with that agent, we consider that agent A held his/her opinion more firmly. Similarly, the agent B held his/her opinion more firmly, too. For the case of $I_A(t) < 0$ and $I_B(t) > 0$ with $D_{AB} < 0$, the result is same. This result shows that the dialogue of people who do not trust each other never leads to an agreement.

However, the calculation result that the opinions of both parties are infinitely distant here is not realistic either. It seems that it will not be affected by widely separated opinions. Therefore, calculation using the theory of this paper is as shown in Fig. 6. In the calculation based on the modified theory, the two who repelled with different opinion from each other, after being separated to some extent, become parallel lines and do not diverge. Such behavior like Fig. 6 shown using the present theory is more realistic.

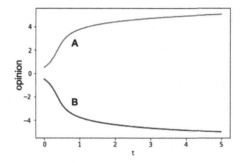

Fig. 6. Calculation result for N = 2. Adv = 0.5, $D_{AB} = -2.0$, $D_{BA} = -2.0$. The initial value is $I_A(0) = 0.005$, $I_B(0) = 0.2$.

3.2   Opinion Dynamics for Three Agents

Next, calculations for the case of three people are shown. The agent A has a positive opinion, the agent B has a negative opinion, and the third agent C has a slightly positive opinion. In the calculation by the previous model [20], even if A is positive opinion and B is opposed as a negative opinion, the phenomenon that A and B gets close to each other due to existence of a third party's C who receives strong trust from both sides is reported.

Let's try the same calculation here as well. Calculation settings are as follows. A is a positive opinion and B is a negative opinion. A and B do not trust each other. On the other hand, C is a slightly positive opinion, but both A and B have great confidence in C. We show in Figs. 7, 8, 9 and 10 for D_{AC} and D_{BC} are 1.5, 2.0, 2.1 and 2.5, respectively. Looking at the above calculation results, it is clear that A and B are likely to compromise due to the influence of intermediary C. And it also shows that it depends greatly on C's personal appeal of how much C is trusted from the surroundings. A and B's opinion are getting closer by brokerage of C who is trusted. Thus, the opinion of A and B approach to C's opinion. In other words, C is a mediator with strong political power that can solve conflict. One example of the person C would be the former president of the Republic of South Africa, Nelson Rolihlahla Mandela who instructed the Republic of South Africa so that all peacefully settled.

Figure 10 shows an example calculated for 300 people using the new opinion dynamics theory of the paper. The mutual D_{ij} value between 300 people is decided by homogeneous random number between −1 and 1. As can be seen in the figure, people's opinions are positive and negative, but they are scattered to a certain extent. According to the calculations assuming 300 people, even though people's opinions are distributed uniformly both positively and negatively, it seems that some degree of equilibrium is reached as a whole. Also, as we can see in the calculated distribution, the calculation starting from a uniform distribution of opinion distribution with a small difference spreads to some extent and is in equilibrium, but its final opinion distribution is not uniform. We can find some opinion groups in the calculated distribution.

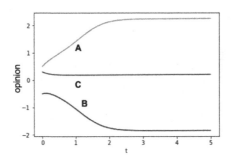

Fig. 7. Calculation result for N = 3. Adv = 0.005, $b = 5$, $\beta = 1.0$, $D_{AB} = -2.0$, $D_{AC} = 3.0$, $D_{BA} = -2.0$, $D_{BC} = 3.0$, $D_{CA} = 1.0$, $D_{CB} = 1.0$. The initial value is $I_A(0) = 0.5$, $I_B(0) = -0.5$, $I_C(0) = 0.2$.

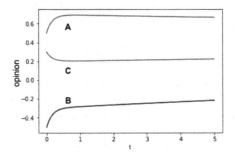

Fig. 8. Calculation result for N = 3. Adv = 0.005, $b = 5$, $\beta = 1.0$, $D_{AB} = -2.0$, $D_{AC} = 4.0$, $D_{BA} = -2.0$, $D_{BC} = 4.0$, $D_{CA} = 1.0$, $D_{CB} = 1.0$. The initial value is $I_A(0) = 0.5$, $I_B(0) = -0.5$, $I_C(0) = 0.2$.

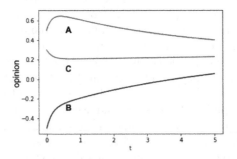

Fig. 9. Calculation result for N = 3. Adv = 0.005, $b = 5$, $\beta = 1.0$, $D_{AB} = -2.0$, $D_{AC} = 4.2$, $D_{BA} = -2.0$, $D_{BC} = 4.2$, $D_{CA} = 1.0$, $D_{CB} = 1.0$. The initial value is $I_A(0) = 0.5$, $I_B(0) = -0.5$, $I_C(0) = 0.2$.

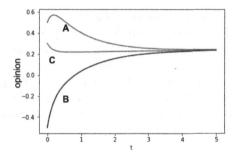

Fig. 10. Calculation result for $N = 3$. $\text{Adv} = 0.005$, $b = 5$, $\beta = 1.0$, $D_{AB} = -2.0$, $D_{AC} = 5.0$, $D_{BA} = -2.0$, $D_{BC} = 5.0$, $D_{CA} = 1.0$, $D_{CB} = 1.0$. The initial value is $I_A(0) = 0.5$, $I_B(0) = -0.5$, $I_C(0) = 0.2$.

The opinion distribution in the figure can be measured on social media [22, 23], such that the present theory can be checked by observations on social media.

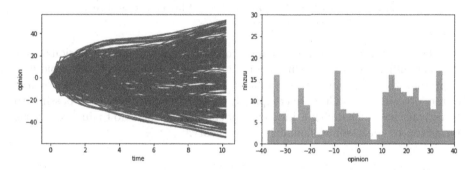

Fig. 11. Calculation result for $N = 300$. The left is the trajectories of opinions. The right is the distribution of opinions at the final of this calculation.

4   Discussion

Looking at the calculation for the case of two people, if there is a relationship of trust with each other, they will compromise as they compromise. In addition, they oppose each other with distrust and rejected their opinions, and follow a parallel line with opinions that are somewhat distant. The same is true for the three people, but it turns out that if there is a mediator between two people with different opinions, there is a possibility that the two will be compromised. In that case, it depends greatly on the competence of intermediaries. It seems that it is the accumulation of these effects that charismatic politicians put together domestic well.

It is shown in Fig. 11 that this theory can be generally calculated by N persons. From these N person's calculations, various aspects of society can be reproduced by calculation. For example, if all the society is connected by trust, it can be expected that consensus building is ready. On the other hand, in the case shown in Fig. 12, it can also calculate that the society will divide. Furthermore, if an attractive charismatic politician appears, it will be possible to predict how such division can be prevented. The charismatic politician can be set in this theory as a person who get strong trust from all other people. Also, when minorities in society are isolated, it will be possible to simulate how isolation can be prevented if we put in our hands.

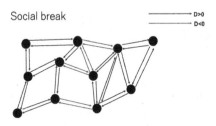

Fig. 12. Schematic illustration of social break of this theory.

The theory presented in this paper makes the opinion one-dimensional, but it is not difficult to extend this theory to multi-dimensional. For example it is clear that the opposing axis of politics and the conflict axis of each country of the football World Cup are completely independent. It is thought possible to investigate how such multiple independent conflicting axes lead to final consensus formation by multi-dimensionalizing this theory.

5   Conclusion

In this research, we presented a theory of opinion dynamics where we consider the opinion as a continuous value. Opinions are represented by real numbers ranging from positive to negative. We introduce "trust" and "distrust" as a coefficient of each person pairs. In addition to the influence of opinion exchanges within each group, we constructed a mathematical model that incorporates external pressure. The theory presented in this paper is modified from the previous model. Using this theory, we can mathematically express many phenomena that can occur in a group in society.

In this new opinion dynamics theory, it is possible to calculate the dynamics of a complicated system mixed with people's trust and suspicion. Simulation of a large number of people is also prepared. As a future prospect, this theory will be useful not only for consensus building but also for serious conflicts in society and the division of society by it and searching for ways to avoid it. In that case, the role of the charismatic leader and the like can be verified from the theory.

Acknowledgement. The author is grateful for fruitful discussion for this theory with Prof. Serge Galam of Institut dÉtudes Politiques de Paris. The author is also grateful for discussions with Dr. Y. Kawahata for this theory. The author also thank Mr. Y. Horio for supporting the coding of the calculation program of N persons.

References

1. French, J.R.P.: A formal theory of social power. Psychol. Rev. **63**, 181–194 (1956)
2. Harary, F.: A criterion for unanimity in French's theory of social power. In: Cartwright, D. (ed.) Studies in Social Power. Institute for Social Research, Ann Arbor (1959)
3. Abelson, R.P.: Mathematical models of the distribution of attitudes under controversy. In: Frederiksen, N., Gulliksen, H. (eds.) Contributions to Mathematical Psychology. Holt, Rinehart, and Winston, New York (1964)
4. De Groot, M.H.: Reaching a consensus. J. Am. Stat. Assoc. **69**, 118–121 (1974)
5. Lehrer, K.: Social consensus and rational agnoiology. Synthese **31**, 141–160 (1975)
6. Chatterjee, S.: Reaching a consensus: some limit theorems. In: Proceedings of International Statistics Institute, pp. 159–164 (1975)
7. Sîrbu, A., Loreto, V., Servedio, V.D.P., Tria, F.: Opinion dynamics: models, extensions and external effects. In: Loreto, V., et al. (eds.) Participatory Sensing, Opinions and Collective Awareness. Understanding Complex Systems. Springer, Cham (2017). https://doi.org/10.1007/978-3-319-25658-0_17
8. Castellano, C., Fortunato, S., Loreto, V.: Statistical physics of social dynamics. Rev. Mod. Phys. **81**, 591–646 (2009)
9. Clifford, P., Sudbury, A.: Biometrika **60**, 581–588 (1973)
10. Holley, R., Liggett, T.: Ann. Probab **3**, 643–663 (1975)
11. Galam, S.: Physica A **238**, 66 (1997)
12. Sznajd-Weron, K., Sznajd, J.: Int. J. Mod. Phys. C **11**, 1157 (2000)
13. Sznajd-Weron, K., Tabiszewski, M., Timpanaro, A.M.: Europhys. Lett. **96**, 48002 (2011)
14. Galam, S.: Application of statistical physics to politics. Phys. A: Stat. Mech. Appl. **274**, 132–139 (1999)
15. Galam, S.: Real space renormalization group and totalitarian paradox of majority rule voting. Phys. A: Stat. Mech. Appl. **285**(1–2), 66–76 (2000)
16. Hegselmann, R., Krause, U.: Opinion dynamics and bounded confidence models, analysis, and simulation. J. Artif. Soc. Soc. Simul. **5** (2002)
17. Deffuant, G., Neau, D., Amblard, F., Weisbuch, G.: Mixing beliefs among interacting agents. Adv. Complex Syst. **3**, 87–98 (2000)
18. Weisbuch, G., Deffuant, G., Amblard, F., Nadal, J.-P.: Meet, Discuss and segregate!. Complexity **7**(3), 55–63 (2002)
19. Abelson, R.P., Bernstein, A.: A computer simulation model of community referendum controversies. Publ. Opin. Q. **27**, 93–122 (1963)
20. Ishii, A., Kawahata, Y.: Opinion dynamics theory for analysis of consensus formation and division of opinion on the internet. In: Proceedings of The 22nd Asia Pacific Symposium on Intelligent and Evolutionary Systems (IES 2018) (in press)
21. Ishii, A., et al.: The 'hit' phenomenon: a mathematical model of human dynamics interactions as a stochastic process. New J. Phys. **14**, 063018 (2012)

22. Bail, C.A., et al.: Exposure of opposing views on social media can increase political polarization. PNAS **115**, 9216–9221 (2018)
23. Sakamoto, T., Takikawa, H.: Cross-national measurement of polarization in political discourse: analyzing floor debate in the US the Japanese legislatures. In: 2017 IEEE International Conference on Big Data (Big Data), pp. 3104–3110. IEEE (2017)

Negotiation Support Systems and Studies (NS3)

A Framework for Gamified Electronic Negotiation Training

Andreas Schmid$^{(\boxtimes)}$ ⓘ and Mareike Schoop ⓘ

Department of Information Systems I, University of Hohenheim,
Stuttgart, Germany
{a_schmid, schoop}@uni-hohenheim.de

Abstract. The continual digitalisation of business processes requires individuals nowadays to learn to negotiate electronically. Negotiation trainings frequently use negotiation support systems (NSSs) to facilitate the development of electronic negotiation skills. Current NSSs offer a rich set of support functions but fail to provide constructive feedback to the learners regarding their negotiation performance, i.e., whether they reached a good agreement or how they can improve. To address this gap, the current paper suggests a novel approach for electronic negotiation trainings by including game elements in an NSS, thereby offering feedback and increasing the motivation and engagement of negotiators. The requirements for the design of such a gamified NSS are based on an integrative literature review on the state of electronic negotiation training, motivation theories, and gamification. Finally, we present a new framework for electronic negotiation training offering constructive feedback and motivational incentives as part of the NSS. Both elements are expected to enhance learners' engagement and improve their learning outcomes when participating in such an electronic negotiation training.

Keywords: Gamification · Negotiation training · e-Learning ·
Negotiation support systems

1 Introduction

Negotiation skills are an important asset in today's business life. Ideally, negotiators possess the required skills to reach optimal outcomes, save transaction costs, and maintain long-lasting relationships with important business partners. Negotiation skills involve communication, analysis and decision-making [1], which can be learned through years of experience or via dedicated negotiation trainings. The field of negotiation trainings rapidly emerged during the 1980s [2] and provoked several research studies until today on how to teach negotiations [3, 4].

At the same time, ICT and the digitalisation of business process rapidly changed business practices. Nowadays, negotiations are often conducted using electronic media, especially using email [5]. Furthermore, the new technological possibilities gave rise to the development of electronic negotiation systems, which support the negotiators at least in their communication or decision-making tasks [6]. In recent years, negotiation support systems (NSSs) as the predominant representatives of electronic negotiation

© Springer Nature Switzerland AG 2019
D. C. Morais et al. (Eds.): GDN 2019, LNBIP 351, pp. 207–222, 2019.
https://doi.org/10.1007/978-3-030-21711-2_16

systems have been frequently used in negotiation trainings to facilitate the development of additional skills for electronic negotiations [7–10]. NSSs offer communication and decision support and might also facilitate conflict management and document management [11]. Human negotiators are still in control of the negotiation process and decide whether to accept or reject an offer [12]. The NSSs employed in these trainings follow a structured, asynchronous bilateral negotiation protocol and include the exchange of textual messages [7, 9, 13]. Representatives of the used systems are e.g. the Negoisst system [11, 12] or Inspire [14].

When participating in a negotiation training facilitated by an NSS, students' development of negotiation skills crucially relies on reflections about one's negotiation process and performance [2, 7]. Reflections are often facilitated "offline" by structured debriefings [7] or journal entries [4]. Obviously, the feedback mechanisms NSSs currently provide are not sufficient to facilitate such reflections. Students require feedback whether they perform well in their negotiations and what they can improve immediately or in subsequent negotiations.

Besides learning from reflections on negotiation simulations, learning outcomes in general are positively influenced by students' motivation and their resulting engagement in learning tasks [15, 16]. As a distal consequence, engagement also positively impacts academic success [15]. Therefore, fostering engagement to improve learning outcomes is one of the key challenges for instructors.

One recent approach to facilitate engagement is gamification, defined as "[...] the use of game design elements in non-game contexts" [17, p. 10]. Gamification makes use of the powerful motivational elements of games whilst not turning the gamified context into a real game. In addition to their motivational power, several standard game elements such as points, badges, leaderboards, or performance graphs also include feedback on the users' actions [18]. Gamification in the education area is especially employed in web-based and asynchronous systems [19]. Reviews on gamification approaches in general or specifically in the education area predominantly report positive effects on engagement [20–23].

We expect a gamified NSS used in negotiation trainings to lead to greater engagement and better learning outcomes. Furthermore, as an intensive exchange of offers leads to more integrative agreements [24], we suggest students to reach better agreements in terms of individual as well as joint utility.

However, gamifying an existing information system requires more than simply adding some game design elements such as points or badges. Literature and experts in the field of gamification highlight the need to analyse and understand the users' needs, psychological processes, and the context of the implementation for the design process [25].

In this sense and as part of a design science research approach [26], this paper follows the explanatory design theory [27] and presents the general requirements that have to be considered for gamifying an NSS in the context of negotiation trainings. In particular, we conducted an integrative literature review [28, 29] focussing on general learning theories and the characteristics of learning to negotiate electronically (cf. Sect. 2) as well as the underlying motivational theories behind gamification and learning (cf. Sect. 3). In Sect. 4, we present and align our findings with a negotiation process model for electronic negotiation systems. We finally present a framework derived from the kernel theories for motivation, learning and negotiation, which forms the basis for the design of such a gamified NSS. The last chapter shortly discusses the results and presents our next steps.

2   State-of-the-Art in Electronic Negotiations Training

Negotiation trainings combine transmission of theoretical knowledge and principles with practical applications to develop negotiation skills, e.g. by engaging students in role plays, case studies, or simulations [2]. Loewenstein and Thompson [3] provide a list of five teaching methods for negotiations: principle learning, trial-and-error learning, observational learning, learning via feedback, and analogy learning. All these methods emphasise the need to integrate theory and practice in negotiation trainings.

Most negotiation trainings follow Kolb's experiential learning methodology [2, 7, 13], which is rooted in the constructivist learning paradigm. Experiential learning is defined as "[…] the process whereby knowledge is created through the transformation of experience" [30, p. 41]. The process includes the following four phases: (1) having an experience, (2) reflective observations on this experience, (3) drawing conclusions from the experience through abstract conceptualisation and (4) active experimentation in new situations. Because most people already possess limited negotiation knowledge from individual experiences like the fixed pie assumption, experiential learning is seen as a key method to challenge existing knowledge and integrate new concepts [7].

Negotiation trainings often integrate NSSs to conduct negotiation simulations [7, 9, 13]. Negotiating with an NSS requires additional knowledge and skills. As social cues such as mimics and gestures cannot be observed via the system, students have to learn to read between the lines and how to build up a positive relationship [7]. They further learn to use the asynchronous mode of the system to their advantage, e.g. making efficient use of more preparation time as well as the use of time pressure tactics. Negotiating with an NSS also requires students to understand and utilise specific support features and gather experience in using such an NSS [31]. The course design for negotiation trainings, therefore, typically promotes face-to-face negotiation skills first, which are a necessary prerequisite before students can develop negotiation skills for electronic scenarios [7, 13].

At the core of electronic negotiation training is the participation of a student in a negotiation simulation. The design of the training and of the negotiation simulations depends on the learning goals that should be achieved. To avoid excessive demands while using a complex NSS and gathering experience with its support functions, previous courses e.g. started with rather simple single-issue negotiations and proceeded with more complex, multi-issue negotiations [7]. During the negotiation process, students perform different actions and requests in the NSS. The NSS provides different immediate feedback mechanisms for the actions, e.g. a utility value supporting the evaluation of offers.

The use of negotiation simulations may either follow a trial-and-error learning approach or the learning via feedback method [3]. Simple participation in a negotiation simulation without reflecting on the negotiation process is not sufficient for in-depth learning and acquisition of negotiation skills [2, 7]. Consequently, the experiential learning phases of reflective observations and abstract conceptualisation are often facilitated by structured debriefings [2, 7] or journal entries [2, 32] after a negotiation simulation has been concluded. However, the feedback supporting the reflection phase in electronic negotiations training is not a part of the NSS itself but provided by the surrounding training course.

Students' participation and engagement in the electronic negotiation training is driven by their motivation and the goals they pursue. If a student is not motivated at all – a state called a motivation [33] – the student will not participate and engage in the electronic negotiation training and acquire any negotiation skills. However, when students are motivated to participate, their motivation may still greatly differ and impact their learning outcomes [34]. We will have a look at motivation and goals in the following section to explain why and how a student engages in certain tasks.

3 Motivation Theories

Motivation is at the core of any action a person performs. If no motivation is present, a person will not carry out a task at all. Fundamental for all motivational theories is the distinction between extrinsic and intrinsic motivation. An intrinsically motivated person performs a task for the inherent satisfaction provided by the task itself [33]. An extrinsically motivated person performs a task to achieve a separable outcome (e.g. rewards) or to avoid negative consequences such as punishments [33].

In the following, relevant motivational theories in the area of learning and gamification will shortly be presented.

3.1 Self-Determination Theory

Self-Determination Theory (SDT) is one of the most discussed theoretical frameworks in gamification research [20]. SDT is based on several empirical studies and investigates social and environmental factors that foster or undermine motivation [35]. According to SDT, an intrinsically motivated behaviour requires three basic psychological needs to be fulfilled, namely the need for competence, autonomy and relatedness [34].

An individual's need for competence can be fostered by different contextual elements that cause feelings of competence for a task [34], e.g. constructive feedback, rewards or optimal challenges. For an intrinsically motivated behaviour, however, it is also necessary that the individual perceives their action as self-determined and autonomous. Tangible rewards have been shown to have an undermining effect on intrinsic motivation, because they shift the perceived locus of causality for an action to be less self-determined [36]. Similar effects have been found for deadlines, threats, directives, or imposed goals [34]. Relatedness, as the third factor, suggests that intrinsic motivation may additionally flourish if individuals have a feeling of being socially related with other individuals or perceive at least a secure relational base.

Intrinsic motivation is the desirable type of motivation, as it results in enhanced performance, self-esteem, greater persistence, creativity, and high-quality learning [33, 34]. However, most of an individual's actions are not intrinsically motivated. Students completing an assignment for a course, which they value as beneficial for their career, still perform the task for an external outcome rather than for the inherent satisfaction of the task. On the other hand, a student completing the assignment because (s)he feels that otherwise (s)he will fail the course is also extrinsically motivated. Although both students are extrinsically motivated, their motivation differs in their relative autonomy [34].

SDT proposes a continuum of four different types of extrinsic motivation, ranging from externally driven motivation (i.e. external regulation, introjected regulation) to internally driven – yet still extrinsic – motivation (i.e. identified regulation, integrated regulation) [34]. Students with an externally regulated locus of causality for a task show less interest and effort than students with a more internal regulation. More autonomous extrinsic motivation positively influences, among other factors, engagement, enjoyment, performance and higher quality learning [34].

Over time, external regulations will often be internalised and integrated by an individual, i.e. external regulations are brought into congruence with the individual's values, such that the motivation for a behaviour may shift from rather external to a more internal driven locus of causality [34]. The process of internalisation is again fostered by the three factors relatedness, competence and autonomy. Relatedness means, that the behaviour is valued or prompted by others whom an individual feels connected to. Individuals also need the relevant skills to succeed in the behaviour. Finally, in the most autonomous form of motivation, the regulation has been internalised, i.e. individuals understand its meaning and fully integrate it with their own goals and values.

3.2   Flow Theory

The notion of flow shares many similarities with intrinsic motivation and focusses on its subjective phenomenon [37]. Flow is also referred to as an optimal experience, in which an individual is fully immersed and focussed on what is currently done, carrying out the activity for inherent satisfaction. Flow experiences can occur with any activity, whether it is playing a video game, doing sports, or writing a scientific paper [38].

Flow can be experienced when an activity is perceived as challenging but attainable by the individual [37]. A balance of individual skills and the challenge is required, which stretches the individual's skills without exceeding them. When skills and the challenge are not balanced, an individual might either face anxiety or boredom. Neither challenges nor skills are objective variables, as they depend on the subjective perspective of the individual.

Besides the aforementioned balance of skills and challenges, flow theory provides several other implications for the design of games, gamified environments and learning environments [37, 39–41]: First, clear goals help directing the attention towards the activity to perform. Second, unambiguous and immediate feedback on the progress towards reaching the goal is required. And third, a sense of control over one's activity facilitates the occurrence of flow.

3.3 Achievement Goal Theory

Achievement goal theory, a theory originating in the late 1970 s, explains the reasons for students' engagement in specific tasks [42]. In general, goals drive performance through four mechanisms: (1) goals direct attention towards goal-relevant activities, (2) they have an energizing function, i.e. high goals produce greater effort than low goals, (3) they lead to greater persistence and (4) support the use of existing or the discovery of new task-related knowledge and strategies [43].

The core of the theory is formed by the distinction between mastery goals and performance goals [44]. In a mastery goal setting, students focus on the development of competence to successfully accomplish a task. Students following performance goals focus on demonstrating their ability for a task in comparison to others, i.e. they attempt to perform better than others. Within this distinction, mastery goals demonstrated several positive effects, including deep-process learning and self-regulated learning strategies.

Due to mixed results on the effects of performance goals, a further distinction between approach and avoidance was introduced, leading to a 2 × 2 model of achievement goals [44]. Individuals with an avoidance-orientation focus on avoiding not mastering a task (mastery-avoidance) or avoiding performing poorly compared to others (performance-avoidance). Individuals with an approach-orientation possess a more positive attitude towards their goals, i.e. they would like to master a task (mastery-approach) or perform better than others (performance-approach).

A few studies in the domain of negotiation training have investigated the differences between mastery and performance goals. Participants pursuing mastery goals showed greater transfer of learned negotiation skills to new negotiation scenarios than the performance-oriented ones [45, 46].

4   Gamified Electronic Negotiation Training

In the following section, we will integrate previously presented motivational theories and the characteristics of electronic negotiation trainings. First, we will derive the general requirements for a gamified NSS. Finally, based on our results, we present a framework for gamified electronic negotiation training.

4.1   General Requirements

The general requirements in this subsection will be aligned with a five-phase negotiation process model for electronic negotiation systems [47], which was developed to support a wider range of electronic negotiation scenarios than the original model proposed by Kersten [48], that was based on Gulliver's eight-phase model [49]. Each of the subsections represents one phase of the process model, which will shortly be described followed by our findings.

Planning. In the planning phase, negotiators determine their relevant issues to be discussed, aspiration and reservation levels for these issues and the best alternative to a negotiated agreement (BATNA) [50]. They decide about the overall approach (competing or collaborating) and the strategy and tactics used.

From the point of view of negotiation trainings, it is important that negotiators are able to claim value as well as to create value [3]. For a negotiation simulation, this goal should be made explicit, e.g. through the description in a case study. Explicit goals help students to direct their attention and activities towards these goals [43]. At the same time, imposed goals may undermine students' need for autonomy. By defining only a high level goal, i.e. a competing or collaborative approach, the students still have various strategic and tactical choices to reach the same goal. Consequently, students may experiment with various strategies, which should facilitate autonomy and motivation. Therefore, our first requirement is providing **clear goals and freedom for strategic and tactical choices**.

Another challenge for the designers of gamified systems are the effects of the included game design elements. Several game design elements such as quests or leaderboards can be categorised as competitive elements. Including such elements in an NSS may affect negotiation behaviour in the planning phase and in the subsequent phases. In particular, for negotiators that should learn to claim value as well as to create value this may lead to an unbalanced focus on competitive strategies. Consequently, our second requirement is to **balance competitive and cooperative game design elements**.

Agenda Setting and Exploring the Field. In the next step of the negotiation, the negotiators discuss about the negotiation issues and their meanings [47]. As a result, issues may be revised, added or deleted. Eventually, preferences and strategies have to be adapted.

Among other factors, the complexity and difficulty of a negotiation depends on the number of negotiation issues. Single-issue negotiations are quite easy to handle, i.e. they do not necessarily require any decision support in the NSS. Especially for multi-issue negotiations, NSSs offer a utility value to evaluate offers [11]. The cognitive burden for new NSS users is still very high. Therefore, and in-line with previous research [7], we propose to start with simple single-issue negotiations and add more complexity through multiple issues in subsequent negotiations.

Further levels of complexity and difficulty may include variation for the zone of possible agreements or more competitive and/or aggressive negotiation behaviour of the negotiation partner [51]. Our third requirement can be summarised with **providing increasing and optimal challenges**.

The provision of increasing challenges can be argued for from several perspectives: As a first step towards electronic negotiations, students may need to develop required skills to learn reading between the lines due to the absence of social cues. They further need to explore the system, get used to its support features and gather experience using an NSS. Imposing additional burdens through complex negotiation cases at the beginning might lead to excessive demands. In terms of flow theory, this unbalanced relationship between challenges and skills may lead to anxiety [37]. Furthermore, SDT highlights the necessity for optimal challenges so students need for competence can be facilitated. When students successfully accomplish the provided challenges, they will feel more competent and intrinsic motivation is more likely to occur.

Exchanging Offers and Arguments. When the negotiation agenda has been defined, the negotiators start exchanging offers and arguments. The phase is characterised by a continual information exchange regarding issue preferences and priorities [47].

Particularly for this phase, NSSs provide a rich set of support functionalities for communication, decision, conflict and document support. It is critical to remember, that these system functionalities only support the negotiator, i.e. the negotiator is always in control of the negotiation process [12]. Therefore, the use of support functionalities relies either on the user's recognition of their benefits or on the user's need for external help [52]. Consequently, support functionalities are sometimes not used, leading to less efficient negotiations.

As a potential solution, a gamified NSS should offer **incentives to use support functions**. Gamification may not only foster motivation and engagement, but can effectively change user behaviour [53]. When students face these extrinsic incentives first, they might perceive their behaviour for using the support functions as externally regulated [34]. They use support functions not as a result of own beliefs, but to fulfil an external demand. If, however, the students gained experience using the functions and finally recognised their benefits [52], their motivation to use these functions will shift towards a more self-determined types of motivation. In particular, SDT proposes an internalisation process, where the regulated behaviour is positively valued and considered as personally important by the student [34]. Students will then experience greater autonomy in using the support functions. An intensified use of support functions should lead to more efficient outcomes in the negotiation simulations.

Reaching an Agreement. After an intensive exchange of offers, the negotiators may realise that they have successfully elaborated areas for an agreement [47]. They develop joint proposals to settle an agreement. Alternatively, if both parties realise that there is no zone of possible agreements, they may decide to leave the negotiation without a deal.

Especially novice negotiators tend to accept even bad and inefficient agreements, because they try to avoid a failing negotiation [3]. When the negotiators have settled such an inefficient agreement, they may engage in a post-settlement negotiation to improve their outcomes [54]. In electronic negotiations, the majority of post-settlement negotiations is rejected, leaving the negotiators with their initially negotiated, inefficient agreement [55]. Nevertheless, a post-settlement negotiation can be beneficial for complex negotiations where inefficient agreements are more likely to occur [56].

Our intention is definitely not to scrutinise post-settlement negotiations, but to prevent students from accepting a bad agreement because they fear failing. Students must learn that negotiations can also end unsuccessfully, i.e. that no agreement is better than a bad agreement with respect to their BATNA. Negotiators avoiding a failing negotiation show the same behavioural patterns as described in the mastery-avoidance goal-setting [44]. Instead, it would be beneficial for students to follow a mastery-approach goal, where the negotiation simulation receives a positive valence for skill development.

Previous research on the development of negotiation skills also demonstrated positive effects of mastery goals compared to performance goals. Students participating in a mastery-oriented training showed greater transfer of negotiation skills in stressful negotiations than participants in a performance-oriented training [45]. Similarly, another study revealed that students in a performance-oriented setting were less successful in transferring negotiation skills learned in one negotiation scenario to a different negotiation scenario than their mastery-oriented colleagues [46]. When students were faced with an identical negotiation scenario again, there were no differences between the two groups. These results are in-line with other studies on learning (see [44] for a summary), indicating that mastery goals facilitate in-depth learning. Consequently, our next requirement can be summarised with **providing a mastery-approach goal-oriented setting.**

Concluding the Negotiation. If the negotiators have successfully worked towards an agreement, they finally conclude the negotiation [47]. The agreement is evaluated and might consider further improvements. In business negotiations, the negotiators settle their agreement in a contract [12].

To evaluate the achievement of the defined goals and support the reflection phase of the experiential learning cycle, students require feedback after the negotiation was concluded. **Feedback should be provided in a constructive manner**, highlighting positive actions and providing insights on improvements for future tasks. Similarly, negotiators concluding their negotiation unsuccessfully require feedback whether their decision to abort is comprehensible, e.g. because they experienced an impasse or could not find a fair compromise. Constructive feedback framed in a positive manner is especially important as novices tend to be more motivated by positive feedback whereas only experts can be motivated by negative feedback [57]. Furthermore, providing constructive and encouraging feedback supports feelings of competence, which are likely to facilitate intrinsic motivation [34].

The feedback on a failing negotiation or possible improvements for an agreement provides another incentive for the students to repeat a negotiation simulation. In general, gamified learning maintains a positive relationship towards mistakes and failures and allows students to repeat their tasks until they succeed [39, 58]. Reflecting on their previous performance, students will derive their lessons learnt and employ them for future tasks [30]. Indeed, previous research has shown that repeating the same negotiation scenarios enables people to logroll more effectively, leading to more integrative results [59]. Logrolling behaviour also improves across different negotiation scenarios [60]. Furthermore, electronic negotiations provide incentives to experiment with different negotiation approaches [7]. Therefore, and in order to maintain autonomy, we summarise our last requirement with **allowing students to repeat challenges.**

4.2   A Framework for Gamified Electronic Negotiation Training

The previously presented general requirements differ in several ways from the current utilisation of NSSs in electronic negotiation trainings. Starting with the elements of the NSS itself, our new artefact will include game elements which turns the original NSS into a gamified NSS (see Fig. 1). The core of the training is still made up by negotiation simulations, in which the students participate. However, these simulations explicitly form different challenges, i.e. the students will "level up" by starting with rather simple negotiation simulations followed by more complex and difficult negotiation simulations.

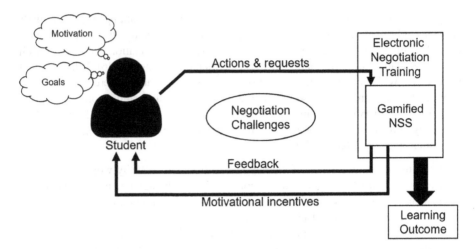

Fig. 1. Framework for gamified electronic negotiation training.

Our gamified NSS explicitly takes into account the motivation and goals that drive students' engagement. The goal of students when participating in a negotiation training should be and probably mostly is mastery-oriented, i.e. becoming *good* negotiators and finding *good* agreements. For novice negotiators it is hard to evaluate whether they actually did a good job in their negotiation, or whether they e.g. failed to find more beneficial, integrative solutions and what they could have done better. Current NSSs do not provide such feedback to the learners. In contrast to state-of-the-art in electronic negotiation training, our new framework includes strengthened feedback provided by the gamified NSS. In addition to the feedback provided during the negotiation process, we extend it with constructive feedback about the negotiation performance, forming the basis for the students' reflections on their negotiation. The constructive feedback replaces course activities such as journal entries [4] or debriefings [7].

Furthermore, the game elements provide motivational incentives to the students. Our requirements especially highlight the satisfaction of the basic psychological needs for competence and autonomy, which according to SDT influence the probability for intrinsic motivation [34]. The incentives to use the support functions of the NSS obviously facilitate extrinsic behaviour but may shift over time to more internally-driven motivation and behaviour. Potential game design elements providing such incentives could be points or badges rewarding desirable and successful use of these support functions [18].

We can expect that the provision of strengthened, constructive feedback and additional motivational incentives will positively impact the students' engagement in the negotiation simulations. Overall, the feedback and increased engagement will hopefully lead to better learning outcomes.

5 Discussion and Outlook

Gamification is an innovative and promising way to foster motivation and engagement of students, and numerous studies in this domain have already been published [20, 22]. Especially qualitative studies often provide a mixed picture regarding the effects of gamified learning [23, 61], therefore understanding the context of the gamification implementation is critical [25].

This paper presents a new approach for electronic negotiation trainings and proposes gamifying NSSs, the systems which are often used within these trainings [7–10]. Following a design science research methodology, we conducted an integrative literature review to derive the general requirements for the design of such a gamified NSS and provide a framework for gamified electronic negotiation training. By synthesising the literature about negotiation training, motivation theories and gamification itself we derived seven general requirements. We cannot guarantee that these requirements and the presented framework are comprehensive, but they provide a well-founded theoretical basis for the design of the artefact.

The core of the gamified electronic negotiation training is students' participation in negotiation simulations. In line with Deterding's method for gameful design [62], we design our gamified NSS around the inherent challenges of the students' activities. Obviously, the inherent challenge in an NSS is to successfully negotiate an agreement with the negotiation partner. The negotiation simulations can have different levels of difficulty, e.g. the number of negotiated issues, the number of negotiation parties participating, varying zones of possible agreements and cooperative or competitive negotiation partners [51]. Challenges are one of the most used game elements in education [23].

The requirements especially highlight the need to provide constructive feedback within the gamified NSS, so the students can reflect on their negotiation performance. Current NSS implementations do not provide such feedback, and the reflection phase is supported by debriefings [7] or journal entries [4]. The requirements further consider and facilitate the students' motivation, another factor that positively impacts students' engagement and their learning outcomes [15].

Employing feedback elements is a common approach in several gamified learning interventions [21, 23]. Feedback is efficient, if it is provided in a positive and constructive way, reflects students' performance and guides students to areas of improvement [39]. To provide feedback, gamification literature in general suggests using points, leaderboards, levels, badges or performance graphs [18, 63], which are frequently used in gamified learning [23]. With respect to the domain of electronic negotiations, further feedback elements should also include domain-specific feedback, e.g. in the form of a Pareto-efficiency graph [64] to display whether integrative potential has been fully exploited.

As the feedback emanates from the NSS itself, our framework for electronic negotiation training does not necessarily require any human negotiation trainer to support the reflection phase. In particular, a negotiation trainer or negotiation expert is only required to design the different negotiation challenges according to the learning goals. Therefore, the gamified NSS does not necessarily have to be embedded within a classical negotiation course but could also be offered as a Massive Open Online Course (MOOC) accessible to everyone interested in learning to negotiate.

Besides the feedback that elements such as points, leaderboards, levels, badges, or performance graphs offer, these elements also provide various motivational mechanisms [18, 65]. Our requirements especially highlight fulfilling the psychological needs for competence and autonomy. However, whether the included game elements will promote extrinsic or intrinsic motivation is highly dependent on contextual factors and the individual student [66]. While this paper primarily focusses on the contextual factors, effects of gamified learning also e.g. differ depending on personality traits and learning styles [67]. Furthermore, undergraduate and postgraduate students have different perceptions on gamified learning interventions [61].

Our next steps, therefore, include careful consideration of game elements that could fulfil our general requirements and might be part of our gamified NSS. Due to the various user characteristics influencing the perceptions and effects of gamification, the risk in designing such a system is user oriented and requires iterative design and formative evaluations [25, 68]. We will report first findings regarding the effects of a gamified NSS on motivation, engagement and learning outcomes as soon as first prototypes have been developed and evaluated.

Acknowledgements. We gratefully acknowledge the funding provided by the Faculty of Business, Economics, and Social Sciences at the University of Hohenheim within the research area "Negotiation Research - Transformation, Technology, Media, and Costs".

References

1. Lewicki, R.J., Barry, B., Saunders, D.M.: Negotiation, 6. edn., internat. ed. McGraw-Hill, Boston (2010)
2. Lewicki, R.J.: Teaching negotiation and dispute resolution in colleges of business: the state of the practice. Negot. J. 13(3), 253–269 (1997). https://doi.org/10.1111/j.1571-9979.1997.tb00131.x
3. Loewenstein, J., Thompson, L.L.: Learning to negotiate: novice and experienced negotiators. In: Thompson, L.L. (ed.) Negotiation Theory and Research, pp. 77–97. Psychology Press, New York (2006)
4. Melzer, P.: A Conceptual Framework for Personalised Learning: Influence Factors, Design, and Support Potentials. Springer, Heidelberg (2018). https://doi.org/10.1007/978-3-658-23095-1
5. Schoop, M., Köhne, F., Staskiewicz, D., et al.: The antecedents of renegotiations in practice —an exploratory analysis. Group Decis. Negot. 17(2), 127–139 (2008). https://doi.org/10.1007/s10726-007-9080-z
6. Ströbel, M., Weinhardt, C.: The Montreal taxonomy for electronic negotiations. Group Decis. Negot. 12(2), 143–164 (2003). https://doi.org/10.1023/A:1023072922126
7. Köszegi, S., Kersten, G.: On-line/off-line: joint negotiation teaching in Montreal and Vienna. Group Decis. Negot. 12(4), 337–345 (2003). https://doi.org/10.1023/A:1024879603397
8. Vetschera, R., Kersten, G., Koeszegi, S.: User assessment of internet-based negotiation support systems: an exploratory study. J. Organ. Comput. Electron. Commer. 16(2), 123–148 (2006). https://doi.org/10.1207/s15327744joce1602_3

9. Melzer, P., Reiser, A., Schoop, M.: Learning to negotiate - the tactical negotiation trainer. In: Mattfeld, D.C., Robra-Bissantz, S. (eds.) Multikonferenz Wirtschaftsinformatik 2012: Tagungsband der MKWI 2012, pp. 1847–1858. GITO; Univ.-Bibl, Berlin (2012)
10. Melzer, P., Schoop, M.: Utilising learning methods in electronic negotiation training. In: Kundisch, D., Suhl., L., Beckmann., L. (eds.) Proceedings of Multikonferenz Wirtschaftsinformatik, pp 776–788 (2014)
11. Schoop, M.: Support of complex electronic negotiations. In: Kilgour, D.M., Eden, C. (eds.) Handbook of Group Decision and Negotiation, pp. 409–423. Springer, Dordrecht (2010). https://doi.org/10.1007/978-90-481-9097-3_24
12. Schoop, M., Jertila, A., List, T.: Negoisst: a negotiation support system for electronic business-to-business negotiations in e-commerce. Data Knowl. Eng. **47**(3), 371–401 (2003). https://doi.org/10.1016/S0169-023X(03)00065-X
13. Melzer, P., Schoop, M.: The effects of personalised negotiation training on learning and performance in electronic negotiations. Group Decis. Negot. **25**(6), 1189–1210 (2016). https://doi.org/10.1007/s10726-016-9481-y
14. Strecker, S., Kersten, G., Kim, J.B., et al.: Electronic negotiation systems: the invite prototype. In: Proceedings of the Collaborative Business MKWI 2006, pp. 315–331. GITO, Potsdam (2006)
15. Kahu, E.R.: Framing student engagement in higher education. Stud. High. Educ. **38**(5), 758–773 (2013). https://doi.org/10.1080/03075079.2011.598505
16. Reschly, A.L., Christenson, S.L.: Jingle, jangle, and conceptual haziness: evolution and future directions of the engagement construct. In: Christenson, S.L., Reschly, A.L., Wylie, C. (eds.) Handbook of Research on Student Engagement, pp. 3–19. Springer, Boston (2012). https://doi.org/10.1007/978-1-4614-2018-7_1
17. Deterding, S., Dixon, D., Khaled, R., et al.: From game design elements to gamefulness: defining gamification. In: Lugmayr, A., Franssila, H., Safran, C., et al. (eds.) Proceedings of the 15th International Academic MindTrek Conference Envisioning Future Media Environments, pp. 9–15. ACM, New York (2011)
18. Sailer, M., Hense, J.U., Mayr, S.K., et al.: How gamification motivates: an experimental study of the effects of specific game design elements on psychological need satisfaction. Comput. Hum. Behav. **69**, 371–380 (2017). https://doi.org/10.1016/j.chb.2016.12.033
19. Buckley, P., Doyle, E.: Gamification and student motivation. Interact. Learn. Environ. **24**(6), 1162–1175 (2016). https://doi.org/10.1080/10494820.2014.964263
20. Seaborn, K., Fels, D.I.: Gamification in theory and action: a survey. Int. J. Hum Comput Stud. **74**, 14–31 (2015). https://doi.org/10.1016/j.ijhcs.2014.09.006
21. Dicheva, D., Dichev, C., Agre, G., et al.: Gamification in education: a systematic mapping study. J. Educ. Technol. Soc. **18**(3), 75–88 (2015)
22. Hamari, J., Koivisto, J., Sarsa, H.: Does gamification work? – A literature review of empirical studies on gamification. In: Proceedings of the 47th Hawaii International Conference on System Sciences (HICSS) 2014, pp. 3025–3034 (2014)
23. Majuri, J., Koivisto, J., Hamari, J.: Gamification of education and learning: a review of empirical literature. In: Koivisto, J., Hamari, J. (eds.) Proceedings of the 2nd International GamiFIN Conference, pp. 11–19 (2018)
24. Gettinger, J., et al.: Impact of and interaction between behavioral and economic decision support in electronic negotiations. In: Hernández, J.E., Zarate, P., Dargam, F., Delibašić, B., Liu, S., Ribeiro, R. (eds.) EWG-DSS 2011. LNBIP, vol. 121, pp. 151–165. Springer, Heidelberg (2012). https://doi.org/10.1007/978-3-642-32191-7_11
25. Morschheuser, B., Hassan, L., Werder, K., et al.: How to design gamification? A method for engineering gamified software. Inf. Softw. Technol. **95**, 219–237 (2018). https://doi.org/10.1016/j.infsof.2017.10.015

26. Hevner, A.R., March, S.T., Park, J., et al.: Design science in information systems research. MIS Q. **28**(1), 75–105 (2004)
27. Baskerville, R., Pries-Heje, J.: Explanatory design theory. Bus. Inf. Syst. Eng. **2**(5), 271–282 (2010). https://doi.org/10.1007/s12599-010-0118-4
28. Torraco, R.J.: Writing integrative literature reviews: guidelines and examples. Hum. Resour. Dev. Rev. **4**(3), 356–367 (2005). https://doi.org/10.1177/1534484305278283
29. Torraco, R.J.: Writing integrative literature reviews: using the past and present to explore the future. Hum. Resour. Dev. Rev. **15**(4), 404–428 (2016). https://doi.org/10.1177/1534484316671606
30. Kolb, D.A.: Experiential Learning: Experience as the Source of Learning and Development. Prentice-Hall, Englewood Cliffs (1984)
31. Melzer, P., Schoop, M.: A conceptual framework for task and tool personalisation in IS education. In: Leidner, D., Ross, J. (eds.) Proceedings of the Thirty Sixth International Conference on Information Systems (ICIS 2015), IS Curriculum and Education, Paper 6 (2015)
32. Melzer, P., Schoop, M.: Personalising the IS Classroom – Insights on Course Design and Implementation. In: Proceedings of the 25th European Conference on Information Systems (ECIS), pp. 1391–1405 (2017)
33. Ryan, R.M., Deci, E.L.: Intrinsic and extrinsic motivations: classic definitions and new directions. Contemp. Educ. Psychol. **25**(1), 54–67 (2000). https://doi.org/10.1006/ceps.1999.1020
34. Ryan, R.M., Deci, E.L.: Self-determination theory and the facilitation of intrinsic motivation, social development, and well-being. Am. Psychol. **55**(1), 68–78 (2000). https://doi.org/10.1037/0003-066X.55.1.68
35. Deci, E.L., Ryan, R.M.: Motivation, personality, and development within embedded social contexts: an overview of self-determination theory. In: Ryan, R.M. (ed.) The Oxford Handbook of Human Motivation, pp. 85–107. Oxford University Press, Oxford (2012)
36. Deci, E.L., Koestner, R., Ryan, R.M.: Extrinsic rewards and intrinsic motivation in education: reconsidered once again. Rev. Educ. Res. **71**(1), 1–27 (2001). https://doi.org/10.3102/00346543071001001
37. Nakamura, J., Csikszentmihalyi, M.: The concept of flow. In: Snyder, C.R. (ed.) Handbook of Positive Psychology, pp. 89–105. Oxford University Press, Oxford (2002)
38. Chen, J.: Flow in games (and everything else). Commun. ACM **50**(4), 31–34 (2007). https://doi.org/10.1145/1232743.1232769
39. Kim, S., Song, K., Lockee, B., et al.: Gamification in Learning and Education: Enjoy Learning Like Gaming. Advances in Game-Based Learning. Springer, Cham (2018). https://doi.org/10.1007/978-3-319-47283-6
40. Sweetser, P., Wyeth, P.: GameFlow: a model for evaluating player enjoyment in games. ACM Comput. Entertain. **3**(3), 3 (2005). https://doi.org/10.1145/1077246.1077253
41. Jackson, S.A.: Flow. In: Ryan, R.M. (ed.) The Oxford Handbook of Human Motivation, pp. 127–140. Oxford University Press, Oxford (2012)
42. Anderman, E.M., Patrick, H.: Achievement goal theory, conceptualization of ability/intelligence, and classroom climate. In: Christenson, S.L., Reschly, A.L., Wylie, C. (eds.) Handbook of Research on Student Engagement, pp. 173–191. Springer, Boston (2012). https://doi.org/10.1007/978-1-4614-2018-7
43. Locke, E.A., Latham, G.P.: Building a practically useful theory of goal setting and task motivation: a 35-year odyssey. Am. Psychol. **57**(9), 705–717 (2002)
44. Murayama, K., Elliot, A.J., Friedman, R.: Achievement goals. In: Ryan, R.M. (ed.) The Oxford Handbook of Human Motivation, pp. 191–207. Oxford University Press, Oxford (2012)

45. Gist, M.E., Stevens, C.K.: Effects of practice conditions and supplemental training method on cognitive learning and interpersonal skill generalization. Organ. Behav. Hum. Decis. Process. **75**(2), 142–169 (1998). https://doi.org/10.1006/obhd.1998.2787

46. Bereby-Meyer, Y., Moran, S., Unger-Aviram, E.: When performance goals deter performance: transfer of skills in integrative negotiations. Organ. Behav. Hum. Decis. Process. **93** (2), 142–154 (2004). https://doi.org/10.1016/j.obhdp.2003.11.001

47. Braun, P., et al.: E-negotiation systems and software agents: methods, models, and applications. In: Gupta, J.N.D., Forgionne, G.A., Mora-Tavarez, M. (eds.) Intelligent Decision-making Support Systems, pp. 271–300. Springer, London (2006). https://doi.org/10.1007/1-84628-231-4_15

48. Kersten, G.E.: Support for group decisions and negotiations: an overview. In: Clímaco, J. (ed.) Multicriteria Analysis, pp. 332–346. Springer, Heidelberg (1997). https://doi.org/10.1007/978-3-642-60667-0_32

49. Gulliver, P.H.: Disputes and Negotiations: A Cross-Cultural Perspective. Studies on Law and Social Control. Academic Press, New York (1979)

50. Fisher, R., Ury, W., Patton, B.: Getting to Yes: Negotiating Agreement Without Giving In. Houghton Mifflin, Boston (1991)

51. Schmid, A., Schoop, M.: Inherent Game Characteristics of Electronic Negotiations. In: UK Academy of Information Systems Proceedings (UKAIS 2018) (2018)

52. Druckman, D., Filzmoser, M., Gettinger, J., et al.: 2.0^2 GeNerationS - avenues for the next generation of pro-active negotiation support. In: Proceedings of Group Decision and Negotiation Conference, vol. I, pp. 82–84 (2012)

53. Blohm, I., Leimeister, J.M.: Gamification: Design of IT-based enhancing services for motivational support and behavioral change. Bus. Inf. Syst. Eng. **5**(4), 275–278 (2013). https://doi.org/10.1007/s12599-013-0273-5

54. Raiffa, H.: Post-settlement settlements. Negot. J. **1**(1), 9–12 (1985). https://doi.org/10.1111/j.1571-9979.1985.tb00286.x

55. Block, C., Gimpel, H., Kersten, G., et al.: Reasons for rejecting pareto-improvements in negotiations. In: Seifert, S., Weinhardt, C., (eds) Proceedings of Group Decision and Negotiation Conference. Universitätsverlag Karlsruhe, Karlsruhe, pp. 243–246 (2006)

56. Gettinger, J., Filzmoser, M., Koeszegi, S.T.: Why can't we settle again? Analysis of factors that influence agreement prospects in the post-settlement phase. J. Bus. Econ. **86**(4), 413–440 (2016). https://doi.org/10.1007/s11573-016-0809-5

57. Fishbach, A., Eyal, T., Finkelstein, S.R.: How positive and negative feedback motivate goal pursuit. Soc. Pers. Psychol. Compass **4**(8), 517–530 (2010). https://doi.org/10.1111/j.1751-9004.2010.00285.x

58. Lee, J.J., Hammer, J.: Gamification in education: what, how, why bother? Acad. Exch. Q. **15** (2), 146–151 (2011)

59. Bazerman, M.H., Magliozzi, T., Neale, M.A.: Integrative bargaining in a competitive market. Organ. Behav. Hum. Decis. Process. **35**(3), 294–313 (1985). https://doi.org/10.1016/0749-5978(85)90026-3

60. Thompson, L.: The influence of experience on negotiation performance. J. Exp. Soc. Psychol. **26**(6), 528–544 (1990). https://doi.org/10.1016/0022-1031(90)90054-P

61. Buckley, P., Doyle, E., Doyle, S.: Game on! Students' perception of gamified learning. J. Educ. Technol. Soc. **20**(3), 1–10 (2017)

62. Deterding, S.: The lens of intrinsic skill atoms: a method for gameful design. Hum. Comput. Interact. **30**(3–4), 294–335 (2015). https://doi.org/10.1080/07370024.2014.993471

63. Mekler, E.D., Brühlmann, F., Tuch, A.N., et al.: Towards understanding the effects of individual gamification elements on intrinsic motivation and performance. Comput. Hum. Behav. **71**, 525–534 (2017). https://doi.org/10.1016/j.chb.2015.08.048

64. Tripp, T.M., Sondak, H.: An evaluation of dependent variables in experimental negotiation studies: impasse rates and pareto efficiency. Organ. Behav. Hum. Decis. Process. **51**(2), 273–295 (1992). https://doi.org/10.1016/0749-5978(92)90014-X
65. Sailer, M., Hense, J., Mandl, H., et al.: Psychological perspectives on motivation through gamification. Interact. Des. Archit. J. **19**, 28–37 (2013)
66. Deci, E.L., Koestner, R., Ryan, R.M.: A meta-analytic review of experiments examining the effects of extrinsic rewards on intrinsic motivation. Psychol. Bull. **125**(6), 627–668 (1999)
67. Buckley, P., Doyle, E.: Individualising gamification: an investigation of the impact of learning styles and personality traits on the efficacy of gamification using a prediction market. Comput. Educ. **106**, 43–55 (2017). https://doi.org/10.1016/j.compedu.2016.11.009
68. Venable, J., Pries-Heje, J., Baskerville, R.: FEDS: a framework for evaluation in design science research. Eur. J. Inf. Syst. **25**(1), 77–89 (2016). https://doi.org/10.1057/ejis.2014.36

Application of Data Mining Methods for Pattern Recognition in Negotiation Support Systems

Muhammed-Fatih Kaya[(⊠)] [iD] and Mareike Schoop [iD]

University of Hohenheim, 70599 Stuttgart, Germany
{Muhammed-Fatih.Kaya, Schoop}@uni-hohenheim.de

Abstract. Data mining methods have long been used to support organisational decision making by analysing organisational data from large databases. The present paper follows this tradition by discussing two different data mining techniques that are being implemented for pattern recognition in Negotiation Support Systems (NSSs), thereby providing process assistance to human negotiators. To this end, data from several international negotiation experiments via NSS Negoisst is used. Consequently, a suitable data representation of the underlying utility data and communication data has to be created for the applicability of data mining. Each generated data type needs individual processing treatments and almost all data mining methods lose their feasibility without a correct data representation as consequence. Once a correct data representation is found, the potential for pattern recognition in electronic negotiation data can be evaluated using descriptive and predictive methods. Whilst Association Rule Discovery is used as a descriptive technique to generate essential sets of strategic association patterns, the Decision Tree is applied as a supervised learning technique for the prediction of classification patterns. The extent to which reliable as well as valuable patterns can be derived from the electronic negotiation data and valuable predictions can be generated is examined in this paper.

Keywords: Negotiation Support Systems · Data mining · Text Mining · Association Rule Discovery · Decision Tree

1 Introduction

Data Mining (DM) is an established research area and has strengthened its pioneering position in data science research in recent years [11]. Nowadays, powerful computer systems are able to analyse large amounts of data with efficient DM algorithms and to generate value-added information [7]. Organisations are interested in extracting useful knowledge from organisational and process data in order to support organisational decision making with valuable patterns, optimisations, and predictive models [6, 38].

The current paper deals with such organisational decision support in the application area of electronic business negotiations – more precisely in Negotiation Support Systems (NSSs) (cf. Sect. 2). In this context, the exploratory application of different DM techniques will show to which extent reliable descriptive or rather predictive

© Springer Nature Switzerland AG 2019
D. C. Morais et al. (Eds.): GDN 2019, LNBIP 351, pp. 223–237, 2019.
https://doi.org/10.1007/978-3-030-21711-2_17

patterns can be detected in interactions of electronic negotiations w.r.t. the common behaviour of negotiation participants using the framework of Knowledge Discovery in Databases (cf. Sect. 3). In order to be able to generate these patterns, a correct data representation has to be found as a prerequisite for DM techniques (cf. Sect. 4). Finding a correct data representation poses the biggest challenge in this work, since heterogeneous types of data (s.a. utility or communication data) are exchanged that require different mining-specific processing as well as transformation steps. As will be discussed later, one method of descriptive character and one of predictive character were selected for the application process of DM and subsequently implemented (cf. Sect. 5). Finally these methods are evaluated w.r.t. the added value of the derived results and the peculiarities of the applied methods. Additionally, the merits of our approach are discussed referring to the overall research goal of pattern detection in NSSs (cf. Sect. 6).

2   The Application Field of Negotiation Support Systems

E-negotiations are not the mere translations of traditional negotiations into the digital realm. Rather, they add a benefit through the application of information and communication technology [33]. In particular, there is at least one rule supporting the decision making, the communication, the document management, and/or the conflict management [33, 36].

A Negotiation Support System (NSS) is designed to advise and support the negotiator during the negotiation process. The most comprehensive NSS is Negoisst [33, 34] which forms our source system for the DM application. Negoisst combines several support functionalities of which communication support and decision support are used in the paper. Negotiators communicate in an asynchronous manner by exchanging semi-structured messages in natural language [31–33]. Each message consists of a content, which is semantically enriched through the combination of natural language terms and the negotiation agenda. A negotiation process ends with a refusal or an acceptance. While a refusal indicates serious disagreement between the parties and terminates the negotiation without an agreement, an acceptance of the latest counter-offer leads to a legally binding contract [33, 34].

The negotiation issues are weighted in terms of preferences. Each issue has a best case and a worst case, defining the range of possible agreement. Based on that information, Negoisst computes a utility function using the hybrid conjoint method [16]. Each (counter-) offer exchanged is then rated with an individual utility value. The joint utility (i.e. the sum of the individual utilities) above 100% shows the agreement to be integrative or win-win as the negotiation pie was enlarged. The fairness of an agreement is measured by the contract imbalance, i.e. the difference between the partners' individual utilities. The lower the difference, the fairer the agreement.

To reach an agreement, several (counter-) offers are exchanged so that utility data and textual communication data are generated during the negotiation process. Negoisst has been employed in university teaching for the past 17 years. As a basis for our research, seven international negotiation experiments via Negoisst (6159 exchanged negotiations messages) were used as source data in order to find indicators for the success or failure of a negotiation. To this end, these large data sets are algorithmically processed using DM methods. As we deal with raw data, which is often unstructured and offers little or no meaning, the pressing question is which process steps have to be conducted to arrive at value-adding findings.

3   The Implementation of KDD Using Data Mining

The methodological basis of the Knowledge Discovery Process (KDD) process model represents an elemental framework to conduct DM methods effectively. It consists of several steps, which can be carried out in a sequential or in an iterative way [8, 15] (Fig. 1).

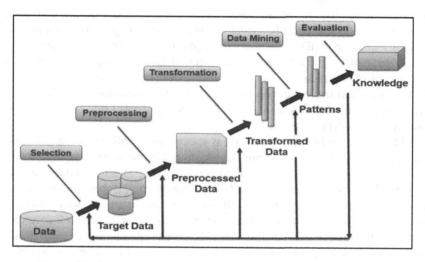

Fig. 1. The knowledge discovery process according to Fayyad et al. [15]

It is rarely the case that the existing data can be transferred directly as input to the DM methods. First, the available data is spotted and selectively transferred to a database, which in our case has already been done by determining the Negoisst source system and the respective data extraction. Based on this data, the selection of suitable data sets takes place (in our case a set of Negoisst experiments). In the following step, the preparation of data for the treatment of missing or impure data takes place. Here, the selected basic data record is cleaned of erroneous data and the respective correction of conflicting values is made.

Preparing the dataset for DM is the most time-consuming part of the overall process. As explained, Negoisst stores utility data and the corresponding textual negotiation message for each exchanged offer. During the transformation process, a valid data representation must be found for both data types so that the DM techniques can be performed without complications. Different data types require different mining-specific processing and transformation steps. Whilst metric utility data might not be comparable across all negotiation experiments, unstructured textual data requires suitable processing steps to be found that enable a valid quantification. After selecting, preprocessing, and transforming the relevant data, it is finally possible to start with active DM.

In general, DM techniques are divided into descriptive techniques and predictive techniques [22]. Selecting the right DM method requires a precise formulation of the problem and the respective analytical objective should be available [14].

In this paper, the Association Rule Discovery (ARD) is selected as a descriptive technique to detect behavioural indicators for acceptance or rejection of the negotiation. ARD measures the strength of coexistence between considered attribute values in large databases that have to meet certain quality measures [2]. The fundamental goal is the determination of Boolean existence connections in the form of if-then relationship patterns $(A \rightarrow B)$ based on the underlying data [3]. Consequently, an association rule consists of a premise A and a consequence B. As we focus on the derivation of systematic patterns that describe the natural behaviour of negotiation participants, the previously described utility values of exchanged offers and their textual messages are used as data input. To check the confidentiality of the rules, support (i.e. the relative frequency of examples in the record) and confidence (i.e. the relative frequency of the examples in which the association rule is correct) are considered as quality measures [41]. Both quality measures are calculated for each generated rule using the "Apriori algorithm" as an iterative method.

Additionally, the Decision Tree (DT) is selected as a predictive technique which is used as a supervised learning method in order to ensure the subsequent interpretability of the results [29]. Through the use of top-down-based divide-and-conquer algorithms, the learning problem is trained based on a number of independent instances and subsequently presented as a DT [39]. Starting from the root node, which reflects the entire basic data set, the data is gradually divided into at least two subsets according to a certain distribution criterion at the nodes [6]. Subsequently, for each generated subset, a child node is again generated. This node marks the next partition criterion [9]. All process steps are carried out iteratively for each generated child node until all training objects (offers in our case) are assigned to a specific target class [18]. The target classes are located on the leaf level of the DT and are represented by the leaf nodes.

To maintain the trustworthiness in addition to the compactness of the tree, the Gini index [30] is considered. Furthermore, pruning algorithms were integrated to counteract the overfitting problem [12, 25]. To measure the overall validity, we use accuracy which is based on precision (i.e. the number of correct predictions of the respective target class divided by the sum of correct and false predictions made for the same target class) and recall (i.e. the number of correct predictions relating to the respective target class divided by the number of tuples belonging to the target class). In summary, both DM methods are performed independently using the KDD and therefore require an individual data preparation process.

4 Data Processing

As described in Sect. 2, each of the (counter-) offers generates utility as well as textual communication data. Hence, both data types must be treated differently in the context of data processing.

4.1 Processing Textual Negotiation Messages Using LIWC

The fact that textual content is in an unstructured form involves considerable effort in terms of pre-processing. Text Mining is a DM application in which the input data may be in the form of written documents, messages etc., which will then be quantified in the pre-processing step [23].

For the targeted analysis of unstructured negotiation communication data, the software "Linguistic Inquiry and Word Count" (LIWC) was used [27]. LIWC has achieved high reliability and acceptance through its extensive analytic potential of text messaging in a variety of applications [20, 21]. LIWC is able to recognise meaningful words from exchanged negotiation messages based on the LIWC dictionary and assign them to the predefined psychological categories [19]. It uses a dictionary approach that looks for more than 2300 words along with their word stem in a text file [17, 24]. For example, the words hatred and ridicule are categorised as negative; the words impressive and hopeful are categorised as positive. After LIWC has analysed all words of the underlying text, a relative value is calculated as the existence frequency of the respective LIWC category for most dimensions. As output, the software generates a list of all considered categories with their respective relative frequency values which in our case refers to the number of total words that appear in the exchanged negotiation message [37]. Within the scope of this paper, only those LIWC variables were taken into account that add value relating to the negotiation environment.

In order to maintain the information richness of the communication data, all communication variables generated from LIWC were mapped to various classes in the following pre-processing step. In this case, most of the LIWC variables require an individual mapping taking the negotiation context into account. The first group of communication variables prepared in this work were the summary variables which are based on algorithms that Pennebaker has published in his research [10]. Initially, each of the variables is represented on a 100 point scale. A high numeric value of the analytic dimension means that the author of a text follows a hierarchical style of communication, whereas a low measure of the variable represents a personal level of interaction. The authentic dimension represents honest and open-minded communication at higher levels. Furthermore, for both variables, a value of 50 represents a balance between the two expressions representing an equilibrium between both polarities so that no dominance can be determined [28].

The consideration of this equilibrium character led to the establishment of a tri-partite target class mapping within our application context. In order not to have to consider the transition of polarity as a punctual boundary (at value 50), a safety factor was introduced at this point. In our application context, this factor was defined as the standard deviation (σ) of ($\pm$) 10 units. Sharp limits can lead to a drastic division, which may have a negative effect on the realistic mapping of the target classes and, therefore, cause a distortion of reality. Consequently, each negotiation message having a value in the interval of [40; 60] was classified as "neutral". In summary, the following target classes were generated for both summary variables: (1) *Authentic* [Guarded; Neutral; Disclosing] and (2) *Analytical* [Hierarchical; Neutral; Personal].

In addition, other behavioural and linguistic categories were used for communication analysis. As described, all these variables are represented in relative values. The use of singular and plural pronouns can be taken as a communicative indicator as a predominant use of "I" rather than "we" can indicate psychological closure [35]. Thus, the variable *Linguistic* was formed, in which messages were assigned to one of the two target classes according to their dominance. If the ratio between the use of singular pronouns and plural pronouns of a negotiation message is balanced, it is assigned to the target class "Equal". Hence, we get the following target classes [Plurality; Equal; Singularity].

To this end, the two most important psychological variables were identified as positive and negative emotions [13]. Negotiation messages are investigated for keywords that suggest emotions of positive or negative polarity. However, a mapping regarding the dominance of negative and positive emotion polarity is not possible here. A descriptive analysis showed that negotiators use much more positive emotions than negative emotions when exchanging messages with Negoisst. A contrasting juxtaposition of the two emotional polarities does not make sense at this point, since the existence of positive polarity would clearly dominate compared to the negative polarity. As a result, both frequency variables had to be treated separately as part of the transformation process. While the *"Negative Emotions"* variable has been checked for existence of negative emotions (target classes: [present; non-existent]), the positive emotion variable has been analysed w.r.t. to its strength (target classes: [high; low]). To determine the limit of existence of high or low occurrence, the method of median split was used [4]. All values that lie below the median are assigned to one target class, whereas the values above are assigned to the other target class.

The remaining LIWC variables are all those frequency variables that were examined for existence or absence as part of our transformation process. These variables are *Risk, Anxiety, Power, Anger* and *Sadness*. Each of the enumerated variables was assigned to one of the two target classes: [present; non-existent]. The fundamental reason for choosing this transformation technique was the rare occurrence of these variables in the negotiation messages. Consequently, it does not make sense to divide these variables into e.g. high/low intensity classes.

4.2 Consistent Utility Values

The process of preparing metric utility data triggers other challenges in comparison to unstructured communication data. Here, the biggest challenge is to ensure the comparability of utility values across all considered negotiation experiments. As described in Sect. 2, seven negotiation experiments conducted via Negoisst form the elementary database for this work. Each of these negotiation experiments deals with different case studies so that each experiment is based on different preference structures. These individual preferences affect the individual utility functions and thus the joint utility value.

Hence, we need a normalisation process that allows the comparability and an equal treatment of individual and joint utility values. As an initial step in the normalisation process, descriptive analyses were performed to determine the minima, maxima, and span of each individual sender utility and receiver utility. The reason for the examination of the span is the determination of the interval range, which depends on the respective negotiation experiment and varies from experiment to experiment. Trivially, the value for the span depends on the minimum and maximum values or how far the index of the minimum value is shifted. This index shift must be balanced at this point in order to obtain consistent data. In order to compensate these individual shift intervals and to be able to carry out an automated normalisation of the individual utility functions, we have used a proprietary developed formula as part of our scientific work:

$$rel\_Indiv\_Utility_{t,i} = \left(\frac{input\_Indiv\_Utility_t}{MinMaxRange_i}\right) - \left(\frac{MinRange_i}{MinMaxRange_i}\right)$$

For each data tuple t of the negotiation experiment i, a respective relative individual utility value $rel\_IndivUtility_{t,i}$ is to be generated in each case. This relative value ensures the comparability across all negotiation experiments. Negoisst generates a sender and a receiver utility. Consequently, the introduced formula is used to normalise both individual utility values. In the numerator of the first term, the individual input value $input\_Indiv\_Utility_t$, which is generated by the system for each submitted offer and is to be normalised, is taken into account for each tuple t. In this regard, the input utility value is normalised with the respective span of the experiment $MinMaxRange_i$. The numerator of the second term contains the minimum value of the respective experiment $MinRange_i$, which is divided by the span of the considered experiment $MinMaxRange_i$, just like the first term. The subtraction is necessary to balance the index shift individually for each of the seven experiments. This critical point is the biggest obstacle for ensuring a consistent comparability of utility values over all given experiments.

Note that a relative individual utility of 0 does not mean the absence of the utility regarding a given offer but rather the minimal utility that can be achieved in the negotiation experiment, whereas a utility of 1 reflects the maximal possible utility.

5 Results

Whether in the selection or cleansing of data, the use of software tools in the context of data science remains an elementary component. We use RapidMiner (RM) for the process-related application of the descriptive ARD and predictive DT. It is a tool with high performance for data pre-processing, optimisation, statistical modelling, DM, and machine learning [5, 40]. In our case, RM required additional pre-processing steps for the automated generation of association rules. In this regard, all target classes, which should be considered in the generation of rule patterns, had to be mapped into a binary matrix representation. While communication-oriented classes could be mapped directly into a binary representation, metric utility data had to be firstly represented as intervals.

Since the number of accepted negotiations clearly dominates the number of failed negotiations and thus can negatively influence the analysis process, the DM methods were performed, beyond the total data set of 6159 negotiation messages, on a balanced record, i.e. on an equalised number of failed and accepted negotiations.

5.1 Association Rule Discovery

As described in the previous section, ARD was performed on different subsets in order to obtain a variety of value-added association descriptions. Initially, it was carried out based on the entire negotiation data which showed that unfortunately only communication-oriented rules could be determined.

No.	Premises	Conclusion	Support	Confiden... ↓
92	Power_Present, Negemo_Present	FinalReject	0.344	0.594
90	Negemo_Present	FinalReject	0.360	0.591
64	Negemo_Present	Power_Present, FinalReject	0.344	0.565
No.	Premises	Conclusion	Support	Confidence ↓
131	Anger_Non-existent, Sad_Non-existent	FinalAccept	0.367	0.621
129	Anger_Non-existent, Singular	FinalAccept	0.334	0.619
125	Anx_Non-existent, Sad_Non-existent	FinalAccept	0.349	0.616
117	Sad_Non-existent	FinalAccept	0.409	0.607
112	Power_Present, Sad_Non-existent	FinalAccept	0.349	0.605

Fig. 2. Association patterns – Balanced datasets

As a result, the ARD was carried out for the balanced negotiation record. In fact, this time rules regarding accepted and failed negotiations are generated, even if the rule combinations do not offer a high reliability. As shown in Fig. 2, all confidence values that refer to an acceptance or rejection of a negotiation are in a confidence interval of 0.565 to 0.621. The first rule recognises that the use of power expressions and the presence of negative emotions can provide initial clues to the failure of negotiations. In addition, this finding is supported by the two subsequent rules (rules 90 and 64) where the sole existence of negative emotions gives an indication of negotiation failure. However, the confidence values of the target class *FinalAccept* are higher with a range from 0.605 to 0.621. This refers to a more probable occurrence of the rules regarding the target class *FinalAccept*. In this context, the absence of anger and sadness (131), the absence of anger using a singular-embossed style of speech (129) and the absence of fear and sadness (125) are identified as indicators for accepted negotiations. In addition, the ARD shows that the use of linguistic power constructs in electronic negotiations does not necessarily lead to the failure of the negotiation.

No.	Premises	Conclusion	Support	Confidence ↓
349	Power_Present	FinalReject	0.914	1
352	Negemo_Present	FinalReject	0.724	1
354	Guarded	FinalReject	0.597	1

Fig. 3. Association patterns – Final Reject

In order to refine the search for association rules in relation to the final outcome of the negotiation, we divide the negotiation data set into two further subsets. In this regard, data partitions are created that contain only negotiation messages from failed negotiations and once negotiation messages from accepted negotiations. Figure 3 shows the most reliable association rules with regard to rejected negotiations. With a high degree of support (0.914), the expression of power leads to the failure of a negotiation in the context of our application leads. In addition, while the use of negative emotions results in the rejection of a negotiation, the use of distanced communication stylistics is recognised as another character for negotiation failure.

No.	Premises	Conclusion	Support	Confidence ↓
480	Power_Present	FinalAccept	0.899	1
481	Anger_Non-existent	FinalAccept	0.863	1
482	Anx_Non-existent	FinalAccept	0.784	1
483	Sad_Non-existent	FinalAccept	0.670	1
484	Singular	FinalAccept	0.649	1
492	Power_Present, Anger_Non-existent	FinalAccept	0.765	1
493	Power_Present, Anx_Non-existent	FinalAccept	0.688	1
494	Power_Present, Sad_Non-existent	FinalAccept	0.579	1

Fig. 4. Association patterns – Final Accept

Figure 4 represents only those rules that are based on accepted messages. Interestingly, the use of the communication-oriented power construct (*Power_Present*) also appears as the strongest premise here (rule 480). However, it leads to a successful conclusion of negotiations and is specified only by taking into consideration of further rules. In this regard, the association rules 492 and 494 show that the existence of power expressions leads to the successful completion of negotiations when emotional influences such as anger, fear and sadness are avoided in the negotiation.

5.2 Decision Tree

As described in Sect. 3, the DT is a supervised learning method and consequently targets a tree with maximum predictive validity which is measured using the Accuracy value. Therefore, the balanced negotiation record is taken into account in order to generate a meaningful DT, because a non-interpretable tree representation with low performance resulted for the target class *Final Reject* when DT was applied to the overall negotiation data set using variations of pre-pruning methods. Nonetheless, pre-pruning is applicable in case of balanced data. Furthermore with detailed analysis of various performance vectors, it can be seen that supervised learning performs better. Through iterative adaptations of different parameter constellations, the performance vector was gradually approximated to its maximum. The model represents a solid basis for supervised learning with recall and precision values above 75%, (see Fig. 5). In addition, it should be noted at this point that the classification process of the DT learns much better for the target class *FinalReject* (recall at the rate of 82.40% and precision at the rate of 77.10%). The precision level of *FinalAccept* remains stable at around 81%, whereas the corresponding recall value is at 75.85%.

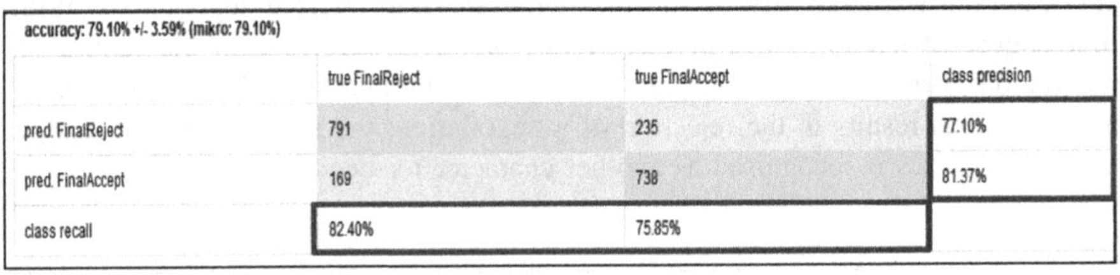

accuracy: 79.10% +/- 3.59% (mikro: 79.10%)			
	true FinalReject	true FinalAccept	class precision
pred. FinalReject	791	235	77.10%
pred. FinalAccept	169	738	81.37%
class recall	82.40%	75.85%	

Fig. 5. Performance vector of the decision tree

The DT includes metric data in addition to communication data and is thus divided into two subtrees at the root node for a relative joint utility (*rel_JointUtility*) of 1.154. Negotiation messages having a value greater than 1.154 are assigned to the left tree, whereas the remaining data tuples smaller than or equal to 1.154 are assigned to the right tree. Consequently, this threshold represents a fundamental split regarding the division of the negotiation outcome.

Furthermore, a large part of negotiation messages are allocated to their target classes after one or two further metric splitting steps. For instance, a closer look at the left subtree shows that a dominant part of the messages are classified as *FinalAccept* (681 negotiation messages) after another utility split at less or equal than 1.367. In this regard, a significant majority of 589 units (86.5%) were correctly predicted as *FinalAccept*. Taking the whole path of the DT into account from the root, this means that a sovereign part of the negotiation, that lies in the joint utility interval of 1.154 to 1.367 and furthermore has an individual receiver utility (*rel_receiverUtility*) of $x > 0.252$, is classified as a *FinalAccept* by the supervised learning method. In return, the splitting criterion *rel_senderUtility* is taken into account as a split variable at the root of the right subtree. A large number of submitted offers (923 tuples) with a relative sender utility over 0.441 are directly assigned to the class *FinalReject*. The quota of correct predictions lies approximately at 77%. Taking the whole path into account, all those negotiating messages are classified with *FinalReject* that have a joint utility of $x > 1.517$ and a relative sender utility of $x > 0.441$. These two described classification paths form the largest groups of target class categorisation.

In summary, 1604 negotiation messages were classified under exclusive consideration of metric split patterns. The remaining 4555 objects were gradually categorised into finer target classes by considering communication-oriented as well as metric decision criteria. For example, the communication variable *Class_Authentic* occurs quite early as a split criterion in the left subtree. While disclosing negotiation messages lead to a successful negotiation according to our learning model, negotiation messages with guarded or neutral communication style are re-examined considering additional metric values. The right subtree includes communication variables in an early phase as well. After examining the metric variable *sender_Utility* described above, it is examined whether the exchanged offers contain emotion-marked signs of sadness in their communication data (*Class_Sad*). If so, they are investigated further in the next tree-levels by an alternation of communication-oriented and metric variables. If sadness does not exist, the negotiation messages are categorised for a successful completion. In summary, communication-oriented variables play an elementary role in both subtrees and they are needed to conduct a detailed classification.

6 Discussion and Outlook

The application of the descriptive ARD as well as the predictive DT are feasible without problems in the context of electronic negotiation data. Consequently, numerous patterns could be generated for the negotiator using the two methods. Nevertheless, specific conspicuousness is observable for each of the two techniques regarding an optimised potential assessment.

In the case of a detailed examination of the ARD results, it can be recognised that only communication-oriented rule associations could be identified. Despite elaborate pre-processing and transformation steps, the underlying "Apriori algorithm" was unable to link communication-oriented data to metric utility values. In this regard, only weak rules were generated that do not provide a valid basis for evaluations and consequently could not be presented in the results chapter. However, this is not a big problem in our case because of the fact that we have considered DT in addition to the ARD as data-mining technique. Nonetheless, numerous significant association rules have been generated. As described before, the algorithm required additional transformation steps. The target classes of the attributes had to be converted to a binary representation. Consequently, all potential target classes were mapped using binary inputs during the process execution. This intermediate step increased the number of input parameters. Therefore, the algorithm had to consider significantly more variables for a potential rule generation. Reflecting this fact reveals the likelihood of potential rule identification in terms of accepted or failed negotiation to be reduced with each additional variable or rather mapped binary target class. Nevertheless, numerous significant association patterns have been extracted in a systematic way, which reflect important behavioural attitudes of the negotiating participants using Negoisst.

The DT compensates the described absence of metric variables in the ARD. In addition to communication-oriented influencing factors, different utility data is included in the classification predictions. The supervised learning process generates a model from which predictions can be made with an overall accuracy of approximately

79.10%. Through iterative adaptations of different parameter constellations, the performance vector of the model was gradually approximated to its maximum. It represents a solid basis for the supervised learning with recall and precision values of more than 75%.

However, a peculiarity can be observed considering the whole DT. As discussed, a big part of the target class classification takes place on the basis of metric utility data. Here, a large part of the negotiation messages is immediately checked after the initial split of the joint utility (at value 1.154) with regard to the individual utility (*rel_senderUtility* or *rel_receiverUtility*). While a large number of *FinalReject* assignments are based on the relative individual utility of the sender in the right subtree (*rel_senderUtility*), the relative utility of the receiver (*rel_receiverUtility*) is taken as the criterion for determining the *FinalAccept* class on the other hand. Thus, a sovereign part of the negotiating messages is classifiable with the sole consideration of utility variables. One possible explanation for this behaviour of the model is the aspect that mathematical relationships between stored utility values are recognised at this point. As a result, systematic utility patterns are generated in the induction process. However, the classification of remaining messages takes place with the inclusion of communication-oriented split variables. Even if their target classes are smaller than those of the initially classified metric splits, they affect and take an important part regarding the classification of the majority. This relates to 74% of the exchanged negotiation messages. Thus, the use of metric as well as communication-oriented variables is indispensable with regard to the induction of the DT.

Even if both DM methods were considered independently, the DT would corroborate the rule combinations from the ARD and add additional value by linking metric and communication-oriented patterns. Furthermore, nearly all detected patterns identified in this work can be confirmed by theories of renowned scientists from the electronic negotiation as well as the classic negotiation environment. This in turn is a clear additional sign for the beneficial usability of the two selected DM techniques within the application field of electronic negotiations.

The generalisability of patterns represents a main limitation of this paper. Despite the combination of a descriptive and a predictive DM technique, it would be wrong to assume the findings as general recommendations for action in electronic negotiations. It should be noted that this work has exclusively used data from the NSS Negoisst. Consequently, the generated patterns represent the negotiating behaviour of Negoisst users. Another limitation is the fact that seven different negotiation experiments from business contexts were used in order to find indicators for the success or the failure of negotiations. The majority of these experiments were based on different case studies with different utility functions. In order to be able to compensate these differences, additional metric-oriented transformation and normalization steps were performed. It would have been desirable to have data coming from the same negotiation case. Potentially, this might lead to further interesting observations regarding the behaviour.

We could show that DM techniques have successfully verified that valuable patterns can be derived from the Negoisst negotiation record. We followed a result-based ex-post approach so that patterns were generated considering the entire negotiation process. However, the approach to the overall negotiation process can be specified by investigating electronic negotiations from a phase-oriented perspective in future research. In doing so, an entire negotiation is divided into several negotiation phases and are subsequently analysed in a more detailed way [26]. As a result, the negotiation behaviour can be analysed step-by-step strategically as well as communicatively. Phase analysis provides a great opportunity to investigate the development of negotiations taking into account logical relationships relating to strategic elements and sequences [1]. They enable the observation of negotiations throughout the communication process and allow negotiators to respond in a more effective manner to the occurrence of unplanned situations through the right action at the right time. This goal can also be enriched or even specified by the use of other DM techniques that generate patterns. The potential of the knowledge discovery process is great and is continuously growing with the help of DM and further machine learning methods.

References

1. Adair, W.L., Brett, J.M.: The negotiation dance: time, culture, and behavioral sequences in negotiation. Organ. Sci. **16**(1), 33–51 (2005). https://doi.org/10.1287/orsc.1040.0102
2. Agrawal, R., Imielinski, T., Swami, A.: Mining association rules between sets of items in large databases. In: Proceedings of the 1993 ACM SIGMOD International Conference on Management of Data, vol. 22, pp. 207–216 (1993)
3. Agrawal, R., Srikant, R.: Fast algorithms for mining association rules in large databases. In: Proceedings of the 20th International Conference on Very Large Data Bases, San Francisco, CA, pp. 487–499 (1994)
4. Aiken, L.S., West, S.G.: Multiple Regression: Testing and Interpreting Interactions. Sage Publication, Newbury Park (1996). https://doi.org/10.1037/0021-9010.84.6.897. (4. Print)
5. Akhtar, F., Hahne, C.: Rapid Miner 5 Operator Reference. Rapid-I GmbH 2012 (2013). http://rapidminer.com/wpcontent/uploads/2013/10/RapidMiner_OperatorReference_en.pdf. Accessed 13 Feb 2015
6. An, A.: Classification methods. In: Encylopedia of Data Warehousing and Mining, pp. 196–201 (2006)
7. Basit, T.: Manual or electronic? The role of coding in qualitative data analysis. Educ. Res. **45** (2), 143–154 (2003). https://doi.org/10.1080/0013188032000133548
8. Brachman, R.J., Anand, T.: The Process of knowledge discovery in databases: a first sketch. In: Proceedings 1994 AAAI Workshop on Knowledge Discovery in Databases, pp. 1–12 (1994)
9. Breiman, L., Friedman, J., Olshen, R., Stone, C.: Classification and Regression Trees. International Group, Wadsworth (1984)
10. Chung, C.K., Pennebaker, J.W.: Linguistic inquiry and word count (LIWC): pronounced "Luke,"... and other useful facts. In: Applied Natural Language Processing: Identification, Investigation and Resolution, pp. 206–229. IGI Global, Hershey (2012). https://doi.org/10.4018/978-1-60960-741-8.ch012
11. Decker, K., Focardi, S.: Technology overview: a report on data mining. Swiss Federal Institute of Technology Technical report CSCS TR-95-02, Zürich (1) (1995)
12. Dietterich, T.: Overfitting and undercomputing in machine learning. ACM Comput. Surv. **27** (3), 326–327 (1995)
13. Druckman, D., Olekalns, M.: Emotions in negotiations. Group Decis. Negot. **17**(1), 1–11 (2007). https://doi.org/10.1145/212094.212114
14. Edelstein, H.: Introduction to Data Mining and Knowledge Discovery, 2nd edn. The Two Crows Corp., Potomac (1998)

15. Fayyad, U.M., Haussler, D., Stolorz, Z.: KDD for science data analysis: issues and examples. In: Proceedings of the Second International Conference on Knowledge Discovery and Data Mining (KDD 1996), Menlo Park-California, pp. 50–56 (1996)
16. Green, P.E.: Hybrid models for conjoint analysis: an expository review. J. Mark. Res. 21(2), 155 (1984)
17. Grinter, R.: Talk to me - foundations for successful individual-group interactions: interact, inform, inspire. In: Conference on Human Factors in Computing Systems, Montreal, Quebec, Canada. Association for Computing Machinery, New York (2006). https://doi.org/10.1145/1124772.1124916
18. Han, J., Kamber, M.: Data Mining: Concepts and Techniques, 3rd edn. Elsevier, Haryana, India, Burlington, MA (2012). https://doi.org/10.1145/565117.565130
19. Hine, M.J., Murphy, S.A., Weber, M., Kersten, G.: The role of emotion and language in dyadic E-negotiations. Group Decis. Negot. 18(3), 193–211 (2009). https://doi.org/10.1007/s10726-008-9151-9
20. Ireland, M.E., Pennebaker, J.W.: Language style matching in writing: synchrony in essays, correspondence, and poetry. J. Pers. Soc. Psychol. 99(3), 549–571 (2010). https://doi.org/10.1037/a0020386
21. Ireland, M.E., Slatcher, R.B., Eastwick, P.W., Scissors, L.E., Finkel, E.J., Pennebaker, J.W.: Language style matching predicts relationship initiation and stability. Psychol. Sci. 22(1), 39–44 (2011). https://doi.org/10.1177/0956797610392928
22. Jain, J., Srivastava, V.: Data mining techniques: a survey paper. Int. J. Res. Eng. Technol. 02, 116–119 (2013)
23. Kotu, V.: Predictive Analytics and Data Mining: Concepts and Practice with Rapidminer, 1st edn. Elsevier, Waltham (2014)
24. Miner, G., Elder, I.V., Hill, T., Nisbet, R., Delen, D.: Practical Text Mining and Statistical Analysis for Non-structured Text Data Applications. Academic Press, Cambridge (2012)
25. Mingers, J.: An empirical comparison of pruning methods for decision tree induction. Mach. Learn. 4(2), 227–243 (1989)
26. Olekalns, M., Brett, J.M., Weingart, L.R.: Phases, transitions and interruptions: modeling processes in multi-party negotiations. Int. J. Confl. Manag. 14, 191–211 (2003)
27. Pennebaker, J.W., Booth, R.E., Francis, M.E.: Linguistic Inquiry and Word Count: LIWC2007–Operator's manual. LIWC.net, Austin, TX (2007)
28. Pennebaker, J.W., Booth, R.J., Boyd, R.L., Francis, M.E.: Linguistic Inquiry and Word Count: LIWC2015: Operator's Manual, pp. 1–22. Pennebaker Conglomerates (2015)
29. Quinlan, J.R.: Induction of decision trees. Mach. Learn. 1(1), 81–106 (1986)
30. Raileanu, L.E., Stoffel, K.: Theoretical comparison between the Gini index and information gain criteria. Ann. Math. Artifi. Intell. 1(41), 77–93 (2004). https://doi.org/10.1023/B:AMAI.0000018580.96245.c6
31. Schoop, M.: The worlds of negotiation. In: Proceedings of the 9th International Working Conference of the Language-Action Perspective on Communication Modelling, LAP, pp. 179–196 (2004)
32. Schoop, M.: A language-action approach to electronic negotiations. Syst. Signs Action 1(1), 62–79 (2005)
33. Schoop, M.: Support of complex electronic negotiations. In: Kilgour, D.M., Eden, C. (eds.) Handbook of Group Decision and Negotiation. AGDN, vol. 4, pp. 409–423. Springer, Heidelberg (2010). https://doi.org/10.1007/978-90-481-9097-3_24
34. Schoop, M., Jertila, A., List, T.: Negoisst: a negotiation support system for electronic business-to-business negotiations in e-commerce. Data Knowl. Eng. 47(3), 371–401 (2003). https://doi.org/10.1016/S0169-023X(03)00065-X

35. Staub, E.: Positive Social Behavior and Morality: Social and Personal Influences. Elsevier Science, Burlington (2013). https://doi.org/10.1016/C2013-0-11528-0
36. Ströbel, M., Weinhardt, C.: The Montreal Taxonomy for Electronic Negotiations Group Decision and Negotiation, vol. 12, no. 2, pp. 143–164 (2003). https://doi.org/10.1023/A: 1023072922126
37. Tausczik, Y.R., Pennebaker, J.W.: The psychological meaning of words: LIWC and computerized text analysis methods. J. Lang. Soc. Psychol. 29(1), 24–54 (2010). https://doi. org/10.1177/0261927X09351676
38. Vercellis, C.: Business Intelligence: Data Mining and Optimization for Decision Making, 1st edn. Willey, New York (2011)
39. Witten, I.H., Frank, E., Hall, M.A.: Data Mining: Practical Machine Learning Tools and Techniques, 3rd edn. Elsevier/Morgan Kaufmann, Amsterdam (2011)
40. Yadav, A.K., Malik, H., Chandel, S.S.: Application of rapid miner in ANN based prediction of solar radiation for assessment of solar energy resource potential of 76 sites in Northwestern India. Renew. Sustain. Energy Rev. 52, 1093–1106 (2015)
41. Zheng, Z., Kohavi, R., Mason, L.: Real world performance of association rule algorithms. In: Provost, F., Srikant, R., Schkolnick, M., Lee, D. (eds.) The Seventh ACM SIGKDD International Conference, San Francisco, California, pp. 401–406 (2001). https://doi.org/10. 1145/502512.502572

Author Index

Printed in the United States
By Bookmasters